BIOPRODUCTS FROM CANADA'S FORESTS

BIOPRODUCTS FROM CANADA'S FORESTS

New Partnerships in the Bioeconomy

by

SUZANNE WETZEL
Natural Resources Canada - Canadian Forest Service,
Sault Ste. Marie,
Ontario, Canada

LUC C. DUCHESNE
Forest Bioproducts Inc.,
Sault Ste. Marie,
Ontario, Canada

Faculté de Foresterie,
Université de Moncton,
New Brunswick, Canada

MICHAEL F. LAPORTE
Natural Resources Canada - Canadian Forest Service,
Sault Ste. Marie,
Ontario, Canada

 Springer

A C.I.P. Catalogue record for this book is available from the Library of Congress.

ISBN-10 1-4020-4991-9 (HB)
ISBN-13 978-1-4020-4991-0 (HB)
ISBN-10 1-4020-4992-7 (e-book)
ISBN-13 978-1-4020-4992-7 (e-book)

Published by Springer,
P.O. Box 17, 3300 AA Dordrecht, The Netherlands.

www.springer.com

Printed on acid-free paper

With the exception of the image of fireweed, all cover images © 2006 JupiterImages Corporation.

Printed in the Netherlands.

Table of Contents

Preface

Highlights and Fast Facts

- This book presents an overview and a vision of bioproducts, specifically as they apply to the forestry sector.
- Bioproducts are commodities that are derived from biomass. Biomass is any type of microbial, plant or organic material (both new and waste) that is available on a renewable or recurring basis.
- Bioproducts include a broad range of commodities with applications to markets such as energy, transportation, chemicals, plastics, foods and food products, pharmaceuticals, nutraceuticals, and various other consumer goods.
- The bioproducts industry is a natural extension of the forestry, non-timber forest products, biotechnology, agricultural, marine, materials and manufacturing industries. Bioproducts from these sectors will form the new bioeoconomy; a market of $100 billion[1].
- The bioeconomy has the potential to stimulate employment and generate wealth in rural and First Nation communities.
- Because the bioproduct industry generates commodities from renewable biomass and wastes instead of fossil fuels, it is often viewed as "green", with the potential of significantly reducing greenhouse gas emissions.
- More investment is needed by industry and governments to stimulate Canada's bioeconomy, to advance biomass as a sustainable alternative to petroleum resources, and to raise awareness of the potential economic, environmental and social benefits of bioproducts.

[1] Unless otherwise stated, all monetary values are reported in Canadian dollars.

In Canada the bioeconomy is expected to generate as much innovation and prosperity as the information technology and energy sectors combined. It will impact most of Canada's economic sectors: energy and transportation, food and agro-food, pharmaceuticals, nutraceuticals, forestry, materials and manufacturing, waste management and a large variety of consumer goods.

The bioeconomy holds promise to wean the Canadian economy from its dependence on fossil fuels as a primary source of energy, as well as platform chemicals in materials and manufacturing, while potentially helping meet Kyoto Protocol commitments on greenhouse gas (GHG) reductions. The bioeconomy also shows potential for reducing the environmental impact of economic activities by increasing the productive use of waste byproducts and developing goods that are biodegradable. Inroads in biotechnology, chemical engineering and an increase in consumer demand for "green" products are added incentives for the expansion of the bioproducts sector in Canada.

Canada is home to 7 percent of the Earth's landmass, 10 percent of its forests, large tracts of arable cropland (over 60 million hectares), 15 percent of the world's fresh water, and is bounded by the world's longest coastline. Coupled with a small population (0.5% of global), this gives Canada a unique resource advantage over most other nations in the world.

Canada's potential for producing biomass feedstocks for energy and other bioproducts is very large and mostly untapped. It is one of the few countries in the world that can rely on supplying a high portion of its own energy needs from biomass, currently meeting about 6 percent of its total needs from this source.

History shows clearly that the availability of cost-effective sources of energy is a critical determinant of economic growth. The world's current decreasing proven reserves and rising costs of non-renewable fossil fuels, along with the need to improve environmental performance, are a challenge for many sectors of the Canadian economy. Commercialization of new and more cost-effective technologies and products in our traditional resource sectors could diversify both markets and energy sources, while boosting rural economies, maintaining our existing industrial infrastructure and stimulating energy independence of northern communities.

In advancing the bioeconomy we are mindful not to create competition with other sectors of the economy. Forests are Canada's greatest source of renewable biomass, and as such are already a primary pillar of the economy, along with the agricultural and marine sectors. Most of the country's biomass production takes place in forests and supports the forest industry as Canada's most productive forests and ecosystems are dedicated to industrial forestry.

In practice, it would make no economic, social or political sense to compete with this industry in support of the bioeconomy. However, there is potentially an estimated 280 million tonnes of forest biomass available for production without interfering with current forestry operations (discussed in Chapter1). A *Primer on Bioproducts*, produced in 2004 by two Canadian Environmental Non-Governmental Organizations, Pollution Probe and the BIOCAP Canada Foundation, concludes that as much as 27 percent of Canada's fossil fuel use could now be supplied from currently underused residue biomass in Canada's forestry and farming operations (crop and animal remains, tree branches, mill waste), from marine residues and from municipal solid waste. A further 25 percent increase in today's tree and crop production could meet an additional 15 percent of the energy demand now being filled by fossil fuels.

Biomass alone isn't sufficient to meet all of Canada's energy needs. Moreover, ecological considerations may restrict biomass availability. Fortunately, its use is compatible with other renewable energy sources. Because it yields liquid fuels that can be stored inexpensively and chemicals that can be used to manufacture durable goods, biomass is highly complementary to hydro, wind and solar energy, which generates instant electrical energy but whose storage is costly.

As new techniques come into commercial application, the cellulose in forest biomass, a potentially much larger and environmentally sustainable resource than agricultural crops such as corn or grain, will become usable as a feedstock for the provision of ethanol and other industrial bioproducts.

In fact, Canada's biomass resources can potentially generate as much as $100 billion annually by providing the feedstock to produce energy, chemicals and materials, including hydrogen, ethanol, methane, pharmaceuticals, nutraceuticals, and bioplastics, as well as a large variety of bio-based consumer goods.

In turn, this new industry has the potential, with participation from all stakeholders, to generate value-added opportunities and prosperity in rural and First Nation communities by drawing on their cultural traditions and creating employment in biomass-rich parts of the country. Governments, in turn, benefit through creation of fiscal revenues and reduced social problems associated with idle workforces. The private sector benefits through new business opportunities either to supplement current operations or as new stand-alone businesses.

A biobased economy can provide a solution to waste management. Energy in the form of carbon from wastes such as pulp residues, municipal carbonaceous material including waste wood and logging residues, can be harnessed to do work before they enter the carbon stream and add to GHG emissions. Furthermore, biomass can provide an alternative energy source for isolated communities that lack the infrastructure and access to fossil fuels, natural gas and electricity. While sensible on a community level, biomass can also provide increasingly important energy security at the national and regional levels.

The writing is on the wall. Car engines will be modified to reduce greenhouse gas and other emissions within the next ten to fifteen years. The large automakers are aggressively researching and developing technologies that will be both practical and inexpensive to replace the gasoline engine. Hydrogen fuel cells are perceived to be an ideal mechanism for energy storage, as its only combustion by-product is water. Currently, natural gas is the most economical source of hydrogen. Forest biomass, however, either as waste byproducts or dedicated biomass, may potentially become the primary source of hydrogen in the future. Indeed, there is evidence that a cord of wood can generate enough hydrogen to propel a hydrogen car for a year. In the more immediate future, there is a need to provide ethanol for blended gasoline, now mandated in several Canadian provinces.

Biomass has the potential to replace many valuable industrial chemicals currently made from non-renewable feedstocks. Platform chemicals are feedstock chemicals with a variety of different uses (see discussion in Chapter 3). Two platform chemicals becoming more important to society are Polylactic Acid (PLA) and Levulinic Acid (LA). Both can be produced from renewable resources like starch crops (corn, potatoes, rice) and cellulosic residues such as logging waste, agricultural waste, and solid municipal waste. This is sometimes referred to as "green chemistry". Both PLA and LA have a wide variety of uses, somewhere between 100-200 potential industrial applications. As a platform chemical, PLA alone currently has market potential in the magnitude of US $10 billion. These two platform bio-chemicals could significantly reduce greenhouse gas emissions, as well as the amount of waste sent to municipal landfills, while increasing profits for rural communities in terms of agriculture and forestry. What about genetically engineered plants that produce plastic monomers? These, too, will play a critical role in the bioeconomy.

Potential problems with accessibility of the resource, competition for woody residues of good quality such as sawdust and shavings by stable secondary industries and ecological questions of removing significant amounts of woody residues, and thereby nutrients, off site will compel innovative techniques to supply a steady feedstock of biomass. Agroforestry (Chapter 4) may well be the key solution for inevitable feedstock shortages. Advances in biotechnology in developing trees with desired traits and planting of marginal farmlands and associated carbon sink potential, ensure agroforestry as a critical provider of goods (both biomass and other bioproducts) of the bioeconomy.

Canada's diverse forest ecosystems have provided nutritious foods to the inhabitants of this land since time immemorial. Some of these foods have become major commodities today while others have largely been supplanted in the market by other crops. Forest foods contribute close to $1 billion in revenue to the Canadian economy (Chapter 5). This potentially lucrative and abundant forest-based food sector could become yet another sustainable source of bioproduct development, including community economic development opportunities, where forest inhabitants can benefit from enhanced access to and production of local forest foods.

Forest bioproducts can also develop in response to changing consumer attitudes and habits. World demand for functional foods and nutraceuticals is estimated at $56 billion and increasing rapidly. Canadian demand is estimated to be in the $1-2 billion range (Chapter 6). This demand is expected to increase with an ageing population that embraces the health benefits of natural products.

The demand for medicinal plants by the pharmaceutical industry is increasing significantly with an estimated 25 percent of prescription drugs in the *US* containing plant extracts or active principles prepared from forest plants. The worldwide industry and potential national and international markets for plant-derived drugs is already worth over US $40 billion (Chapter 7). Since many of our modern medicines have been derived from plants, and since so few plants have been fully investigated, it stands to reason that there are many more beneficial medicines yet to be discovered, with the potential to positively impact the future health of Canadians and the world population at large. Given mutually acceptable sharing of intellectual property, opportunities exist for collaboration with First Nation communities regarding traditional uses of medicinal plants as a means for aiding in the discovery of new forest plant-derived drugs.

More traditional bioproducts (interchangeably also called non-timber forest products, or NTFPs) include ornamental forest bioproducts such as Christmas trees, floral and greenery decorations, and arts and crafts. Although they do not contribute substantially to the gross domestic product, they have significant local impact. The current market for decorative and aesthetic products is roughly $175 million, Christmas trees and salal being the most important (Chapter 8).

Non-consumable values of the forest, generally included in discussions on non-timber values of the forest, such as nature viewing, hunting and fishing continue to be popular activities in Canada. While not tangible bioproducts per se, these activities and values must be considered when assessing the various demands on forested lands. Canadians spend upwards of $11 billion/year on nature-related activities, with the American visitors contributing another $700 million in revenues (Chapter 9). Wild fur trapping, although a shadow of the former industry that led to the exploration of Canada's lands, still contributes $23 million and the fur industry alone provides employment for 68, 000 Canadians.

If forested lands are to succeed as a major support of the bioeconomy, they must be protected from their greatest threats, forest pests. Public pressure is requiring the use of more biological pesticides, which are perceived as safer than the conventional chemical pesticides. Research, aided by the powerful tool of biotechnology, is aggressively targeting the development of biological pesticides, as well as the propagation of trees with selected traits for pest tolerance, increased yield, decreased lignin content, disease tolerance and cold- or drought-hardiness (Chapter 10).

Biotechnology is a critical component of the bioproducts sector and holds the key to enhancing the production of the biomass needed as feedstock for energy production, platform chemicals, and natural fibers. Environmental benefits of these biotechnology applications could include alleviating pressures on old-growth forests through enhanced establishment of agroforestry systems, a reduction in chemical pest-control through the use of target-specific biological pest control products, reduction in chemicals used in pulp and paper processing, and the conversion of formerly wasted residues to create useful products (Chapter 11).

The potential carbon sequestration value of Canadian forests is a popular and valuable short-term non-consumable "bioproduct" (Chapter 12). Forest carbon trading can be used to offset Canada's Kyoto target of reducing greenhouse gas emissions by 6% below 1990 levels and is among the least costly and most immediately available options for offsetting carbon emissions. The expertise certainly exists to manage such forest carbon projects, and emission credits can be generated at low cost which add value and diversify existing investments. Afforestation, reforestation and forest management activities, and the generation of forest carbon credits have the potential to underwrite some of the costs of shifting to more sustainable forest management practices. While government policies are under review, evidence indicates that forest companies, landowners and local governments are ready to begin this revolution. However, investors will, no doubt, continue to be wary of the risks created by complex measurement and verification activities associated with forest carbon management projects in Canada until they are adequately addressed.

Although the Canadian forest industry has a long history of innovation, other financial incentives for research and development will be needed to assist the industrial sector in the implementation of the bioproducts sector in forestry (Chapter 13). This will be critical for the development of technologies and processes that are sustainable from an economic, environmental and social perspective. Strategies must be devised to avoid competition for the resources traditionally used by the forest industry and thus there is a need to integrate the bioproducts sector into the current forestry and forest products sectors. Moreover, there will be a need to generate formal linkages between all sectors of the economy with vested interests in the future of the bioeconomy, to create the new partnerships required for success.

In this book, we paint our vision of the bioeconomy in broad strokes. Our vision entails a highly networked, responsible and competitive forest bioproducts sector in Canada. The book is not intended as a comprehensive treatise of each bioproduct and its market potential. Indeed, one of our most difficult tasks was to decide which topic to address and to what depth. Hence omissions were the result of practical considerations alone and do not reflect value judgment. Indeed, our primary goal was to share our vision so that others could become active members of the bioeconomy.

In practice, strategic directions, along with first order economic, social and environmental aspects of the bioproducts industry, will, in many cases, require additional in-depth analyses, business planning and demonstration research to validate many of our hypotheses and opinions. Advancing the bioeconomy conscientiously, however, will respond to Canadians' unwavering commitment to see sustainable economic growth that enhances traditional resource sectors and rural communities, diversifies energy sources and market opportunities, converts potentially polluting waste into useful products, and ultimately protects their own health, the health of their children, and the health of the environment.

Acknowledgments

This manuscript was researched and written in collaboration with a team of researchers/writers to whom we're deeply indebted for their hard work and patience. K. Brosemer, I. Davidson-Hunt, K. Davidson-Hunt, R. Dredhart, C. Jessup, G. Langlais, B. McKinnon, H. Mitchell, D. Morin, M. O'Flaherty, F. Ortiz, M. Robson & L. Sloane worked arduously at finding information and drafting this manuscript. The manuscript was edited by H. Mitchell and K. Burgess. We're indebted to all for their hard work and thank them for their patience and friendship.

Guidance and help, both much needed and greatly appreciated, were also provided by M. Alvarez, B. Arif, T. Beardmore, D. Booth, W. Beilhartz, A.C. Bonfils, T. Brigham, E. Caldwell, C. Colombo, J. Cunningham, L. Deverno, S. Dominy, J. Eyzaguirre, P. Graham, P. Hall, A. Hopkins, M.F. Lamarche, S. Richard, M. Shaw (the staff of ULERN) and J. Wilson. Many others were consulted and we beg their forgiveness for overlooking their names.

We are thankful to our employers Natural Resources Canada (S. Wetzel and M. Laporte) and the Université de Moncton (L. Duchesne, supported by the Government of New Brunswick) for permitting us to work on this manuscript.

We gratefully acknowledge the funding support of the following:

 **Industry Canada**

Natural Resources Canada

Chapter 1

FORESTS AS A SOURCE OF BIOPRODUCTS

Highlights and Fast Facts

⌇	Canadian forests are a wealth of biodiversity. As many as 500 types of bioproducts derived from forest-based biomass (plants, microbes and animals) are commercially used in Canada today.
⌇	A potential 280 million tonnes of forest-derived biomass are available for production and industrial use without interfering with current forestry operations.
⌇	Biomass utilization is a profitable means to manage residues originating from logging, sawmill and pulp and paper operations.
⌇	Available woody biomass can provide a feedstock to replace fossil fuels for production of various consumer and industrial bioproducts including biofuels, bioplastics, biochemicals and biopesticides.
⌇	One tonne of wood waste biomass contains enough energy to power a car for over six months; via biomass-to-fuel conversion, it may be possible to build a "clean fuel" economy in Canada worth up to $104 billion.[1]
⌇	To benefit from renewable and greenhouse gas-neutral bioenergy and bioproducts, government investment in research and development and the establishment of procurement strategies are needed to augment sustainable forest management practices.

[1] Unless otherwise stated, all monetary values are reported in Canadian dollars.

For over 150 years in Canada, forest resources have been inventoried as merchantable pulpwood and saw logs of commercially desirable tree species. The rest of the forest was perceived as playing a critical ecological role but was rarely the target of specific management. A successful twenty-first century bioproducts industry, however, requires innovative technologies and a new approach to the inventory and monitoring of Canada's forest resources, as well as creating new partnerships in the bioeconomy.

Forest bioproducts can be generated from any organic material found within forest ecosystems or associated forest industries. The capacity to capture value or manufacture new commodities is dependent on the degree of biodiversity present within forest ecosystems. Biomass can originate from timber productive forests, non-timber productive forests, unmanaged or managed stocks, logging wastes, mill wastes or agroforestry systems. The selection of a source of raw material for the bioproducts industry must be the result of careful analysis of a complex combination of economic, environmental, political and social factors. Profitability is a critical issue; however, ecological and social sustainability are equally important success criteria.

1.   FOREST BIODIVERSITY AND BIOPRODUCTS

Forest species that were formally considered not to have commercial value include all microbial, animal or plant species that were not previously exploited by the forest industry. In the emerging bioeconomy, these same species can become a source of raw material for various types of industries, provide supplemental income or become part of major export enterprises. Until now, there has been little attempt at conducting an inventory of this resource, except perhaps for rare and endangered species or those of particular ecological interest, making the current task of evaluating their economic potential quite challenging.

Canadian forests, comprising 10 percent of global forest cover, offer a wealth of biodiversity. Approximately two-thirds of Canada's estimated 140,000 species of plants, animals and microorganisms live in the forest (Natural Resources Canada 2003). Of the species contained in the vast boreal forest, it is estimated that 95% are arthropods and microorganisms, and that 88% of taxa remain unclassified by taxonomists (Zasada et al. 1997). Their genetic diversity includes billions of genes that control basic metabolic functions and confer the ability of all living creatures in Canada's forests to adapt to changing environmental pressures.

All of Canada's forest species, and their inherent genetic diversity, may have commercial values; however, it is nearly impossible to ascribe a monetary value to any species without knowing its specific industrial or commercial application. Nevertheless, as many as 500 types of bioproducts

(Duchesne et al. 2000) derived from plants, microbes and animals are commercially used in Canada. Collectively, these species now contribute close to a billion dollars annually to the Canadian economy, but the potential for harvesting and value-added processing is much greater.

In practice, forest biomass and bioproducts are primarily derived from four main sources: wild stocks from timber productive forests, wild stocks from non-timber productive forests and lands, managed stocks from intensively managed forests, and domesticated stocks from agroforestry ecosystems.

1.1   Wild Stocks from Timber Productive Forests

Each species has its own set of ecological requirements and thus its abundance and distribution is based on its tolerance to, and/or preference of, environmental factors. It is difficult to make general statements about the distribution of individual species. In the case of commercially available wild mushrooms, for example, in the boreal forest of northwestern Ontario, the morels (*Morchella* spp.) is found only in one-year-old post-burn communities (Duchesne and Weber 1993), whereas chanterelle mushrooms (*Cantharellus* spp.) are found in mid-successional forest stands, and pine or matsutake mushrooms (*Tricholoma magnevelare*) are found uniquely in mature (60 year-old-plus) forest stands (A. Chapeskie, per. comm.).

Because bioproducts and timber can be harvested from the same lands, there is potential for conflict if two industries target species with different ecological tolerances. Some bioproducts can benefit from logging disturbance from the timber industry, as in the case of early successional species such as blueberries (*Vaccinium* spp.) in eastern Canada and salal (*Gaultheria shallon* Pursh) in western Canada. At the other end of the spectrum, some species of plants are unique to late successional forests and are sensitive to disturbance. For example, Pacific yew (*Taxus brevifolia*) supplies bark for the production of the cancer drug paclitaxel and occurs as an understorey plant in old-growth forests. As a species, Pacific yew is shade tolerant, slow growing and has a life span of 300 years typically occuring within the understorey of old growth stands of western red cedar (*Thuja plicata*), life span 1000+ years) and western hemlock (*Tsuga heterophylla*), life span 400years, US Department of Agriculture 1999). "Yew habitat is old growth forest. Remove that and yew species will decline" (Lizotte and Knapp 2003).

1.2   Wild Stocks from Non-timber Productive Forests and Lands

The density, abundance and types of bioproducts on non-timber productive lands are similar to those from timber productive forests, but their origin excludes the possibility of direct competition for resources with the timber industry. Non-timber productive forests are not suitable for timber harvesting for various ecological and practical reasons including growth rate, topography, species composition, legislation or land uses incompatible with timber management. As an example, transmission line corridors are a good source of raw materials and can meet the dual goal of controlling vegetation and increasing yields of other products from the land (Saari 1993). Bioproduct management on private land can provide income opportunities for landowners who are not willing or able to manage timber resources (Ihalainen et al. 2002).

1.3   Managed Stocks from Intensively Managed Forests

Intensive forest management offers a wealth of opportunities to augment the social and economic value of forests by increasing the number of commodities extracted. For example, there are many medicinal plants and berry-producing species (Mohammed 1999) that can be co-managed with timber values without conflict between the two industries.

Recently, le Ministère des Ressources naturelles du Québec took a landmark initiative by promoting the intensive management of blueberries along with timber values in alternating strips of land. Local blueberry harvesters who were faced with consumer demands that exceeded the supply prompted this initiative. It was demonstrated that timber outputs within suitable ecosystems were increased by intensively managing fiber (yields increased from 1.8 m^3/ha/yr to 4.7 m^3/ha/yr), which freed up roughly 60% of the land for the intensive production of blueberries (Ministère des Ressources Naturelles 2002). It is expected that by 2007, approximately 5,000 ha of forested land will be available for blueberry production in this combined arrangement (Natural Resources Canada 2003).

Figure 1-1 provides a model that illustrates the potential of bioproducts to enhance forest management productivity through their co-management with timber. According to this model, managing bioproducts in commercial forests can help offset the cost of early plantation silviculture and add to revenues by increasing the number of species harvested from a given parcel of land. This concept is highly pertinent to private lands where tenure is unquestionable. Depending on jurisdiction, this model will have varying applicability on public lands.

1.4   Domesticated Stocks from Agroforestry Ecosystems

Domestication involves actively manipulating the life cycle, growth and environmental factors of organisms, and is the basis of modern agriculture. In practice, there are many instances where the demand for forest-based bioproducts also dictates that they should be domesticated or managed intensively in agroforestry systems.

Cultivation of specialty bioproduct species in agroforestry settings is sometimes undertaken to eliminate the problems associated with finding scant resources from the wild (Ripa 1993), or for reducing pressure on wild stocks of threatened plant species. Such cultivation can help regulate the supply of raw material, a critical factor in all levels of the supply chain. Harvesters need to have access to dependable stocks in order to ensure a steady and predictable income. A secure supply of material is especially critical where large market values are involved, such as wild mushrooms or nutraceuticals, which include highly sought after plants such as ginseng (*Panax* spp.), licorice (*Ligusticum* spp.), cohosh (*Caulophyllum* spp.), ground hemlock (*Taxus canadensis*), ginger (*Asarum* spp.) and Pacific yew (*Taxus brevifolia*) (Wheeler and Hehnen 1993).

Meeting the special needs of consumers is a strong motivation to domesticate a particular type of bioproduct species. The Christmas tree market is one example where consumer demand for products of unique shape and quality cannot be supplied from wild stocks. Across Canada there are several initiatives to domesticate many native forest plants (Montgomery 2000). This concept is also appealing to the agriculture industry as it can offer novel, and often more lucrative, products.

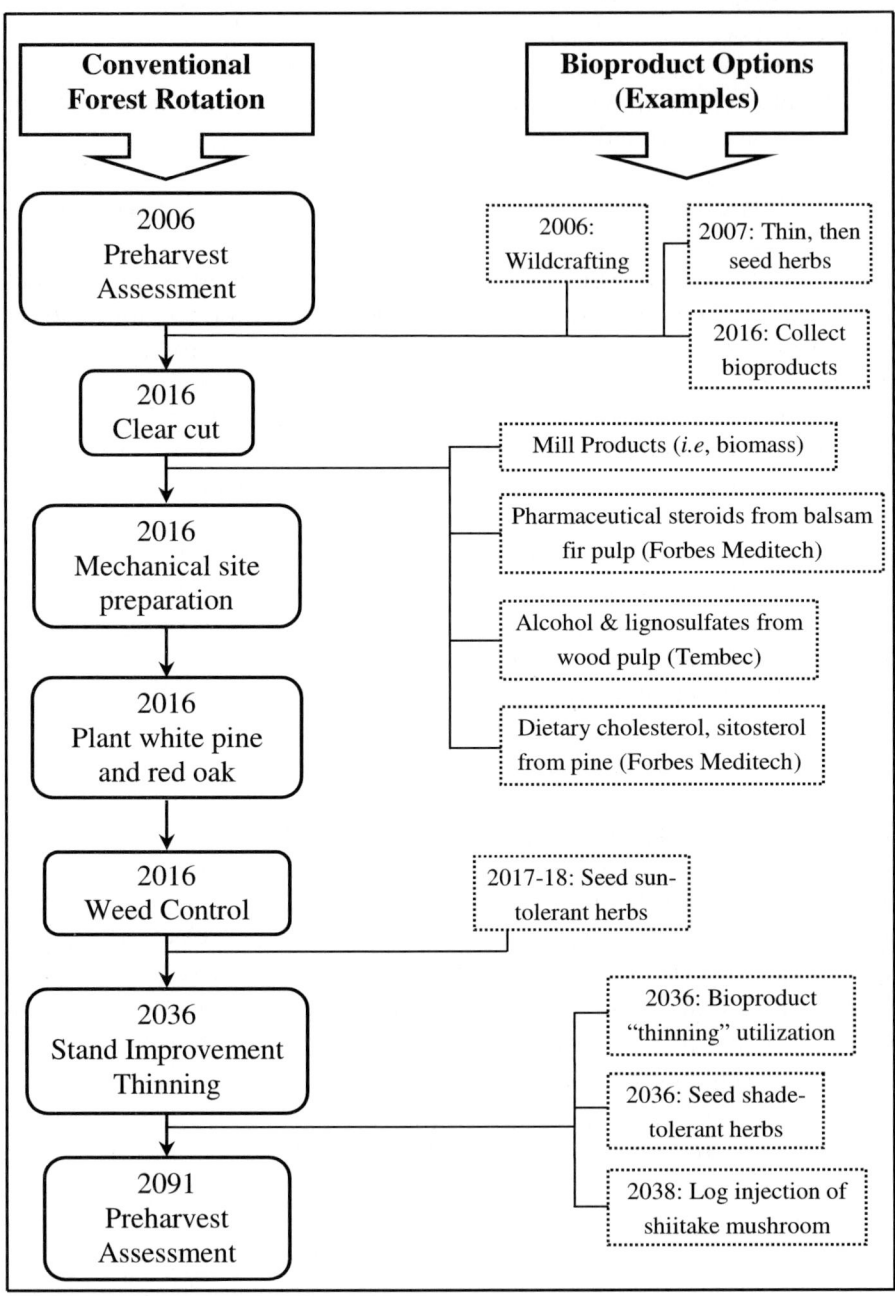

Figure 1-1. Managing Bioproducts in Commercial Forests.

2.   BIOMASS ORIGINATING FROM CANADIAN FORESTS

By far the greatest annual growth of biomass in Canada occurs on forested lands which cover 41.8% (417.6 Mha) of the nation's total 998.5 Mha landmass (Natural Resources Canada 2003; Statistics Canada 2004). The sheer size of this resource provides Canada with a definite "green" advantage. Canada's timber productive forests produce roughly 360 Mm^3 of wood biomass annually, although productivity varies within and among ecozones (Lowe et al. 1996). This renewable source of virgin or residual feedstock is suitable for bioproduct and energy production, the major driver of a bio-based economy. Considering Canada's vast forest resources and the fact that only a small portion of these resources currently contribute to the bioproduct industry, the potential for new enterprise becomes evident.

Wood residues (underused or wasted biomass) are highly attractive from an industrial perspective because of their abundance and low cost. For some bioproducts, residue from harvesting, manufacturing and processing operations is the most efficient resource for commodity production. Not utilized, residue may be burned, left to decompose or landfilled. Given that each of these alternatives carries an environmental cost, conversion of residues into biofuels, or value-added bioproducts, can be an attractive alternative from many perspectives.

2.1   Estimating the Total Biomass Available from Canadian Forests

Table 1-1 summarizes the majority of biomass originating within Canadian forests and is the maximum annual harvest that would be available annually on a sustainable basis. In total, an estimated 278 million tonnes (Mt) of biomass could be made available for conversion to bioproducts. This is biomass originating from Canadian forests (including peat resources), and does not include biomass such as urban yard waste, municipal organic waste and virgin wood from non-timber productive forests.

Table 1-1. Annual Sustainable Canadian Forest Biomass. Economic Value in Canadian Currency and Ratios for Conversion of Biomass to Fuel[1].

Annual Potential Forest Biomass by Type		Commodity[6]							
		Methanol	Ethanol	Bio-oil	Hydrogen[7]	Levulinic Acid	Pellets-Briquettes	Wood Chips	Electricity
Woody Biomass from the Forest 81.50 Mt	B:C[2]	0.33	0.20	0.75	0.048	0.25	1.0	1.0	4.2
	TP[3]	26.90	16.30	61.12	3.912	20.38	81.50	81.50	342.30
	$[4]	30.68	18.59	15.71	12.47	30.30	25.06	9.78	20.54
Logging Residues 75.82 Mt	B:C	0.33	0.20	0.75	0.048	0.25	1.0	1.0	4.2
	TP	25.02	15.16	56.86	3.639	18.96	75.82	75.82	318.44
	$	28.54	17.30	14.62	11.60	28.19	23.32	9.10	19.11
Peat 70.00 Mt	B:C	0.33					1.0		3.5
	TP	23.10					70.00		175.00
	$	26.35					18.22		10.50
Short Rotation Plantations 21.91 Mt	B:C	0.33	0.20	0.75	0.048	0.25	1.0	1.0	4.2
	TP	7.23	4.38	16.43	1.052	5.48	21.91	21.91	195.30
	$	8.25	5.00	4.22	3.352	8.14	6.74	2.63	11.72
Sawmill Residues 19.93 Mt	B:C	0.33	0.20	0.75	0.048	0.25	1.0		4.2
	TP	6.58	3.99	14.95	0.957	4.98	19.93		84.00
	$	7.50	4.55	3.84	3.049	7.41	6.13		5.04
Municipal Waste Wood and Paper 8.04 Mt	B:C	0.33	0.20	0.75	0.048	0.25	1.0	1.0	4.2
	TP	2.65	1.61	6.03	0.386	2.01	3.138[5]	3.138[5]	33.75
	$	3.03	1.83	1.55	1.230	2.99	0.965	0.377	2.02
Paper Mill Sludge 1.55 Mt	B:C	0.33	0.20			0.25	1.0		1.04
	TP	0.52	0.31			0.388	1.550		1.61
	$	0.58	0.35			0.576	0.085		0.097
Commodity Price ($/t)		1140.7	1140.7	257.04	3187.5	1487.0	307.54[8]	120.06	60.00[9]
Total Production[3]		92.00	41.75	155.40	9.946	52.12	273.85	182.37	1150.4
Total Economic Value[4]		104.93	47.62	39.94	31.70	77.61	80.52	21.89	69.03

[1]Calculations for determining available biomass, commodity prices and conversion ratios are found in Appendix A.

[2]Biomass (B) to commodity (C) conversion ratio.

[3]Total production (TP) of commodities in millions of tonnes (Mt) or (mWh for electricity).

[4]Gross annual economic potential at competitive retail prices in billions of Cdn dollars. Some commodities are based on emerging technologies.

[5]The component of waste wood and paper that is wood is estimated to be 3.138 Mt.

[6]Some cellulosic conversion technologies are still being developed and are not yet cost competitive.

[7]Produced from the intermediate conversion of woody biomass to bio-oil.

[8]Peat priced at $260.30/t. Paper mill sludge priced at $55.05/t.

[9]Commodity price in mWh.

To understand the economic potential of the biomass quantified in Table 1-1, refer to Appendix A. This appendix provides the rationale and analysis for the calculations used to determine the biomass-to-commodity conversion ratios and the commodity price assumptions.

The need to define and prove sustainable productivity regarding organic matter or biomass removal from ecosystems is ongoing. Obvious declines in ecosystem productivity are rare and are usually attributed to poor soil management; however, uncertainty will continue to exist until sustainable productivity can be demonstrated conclusively (Powers 1999).

2.2   Forest Biomass as a Feedstock for Bioenergy Production

A potentially lucrative and environmentally benign venture involves the feasibility of using biomass as a renewable source of fuel. For example, one tonne of wood waste contains the equivalent of 18 GJ (Gigajoule 1×10^9 joules) of energy which is sufficient to operate a car for over six months (See Table 1-2).

Table 1-2. Examples of Energy Consumption (GJ) in Canadian Society[1].

Energy Consumption Mode (per year)	Energy Intensity (GJ)	% Change 1990 - 2002
Total Canadian Household Energy Use	1.4×10^9	8.6
One Individual Household - Total Energy Use	116.4	−10.6
- Space Heating	69.1	6.6
- Water Heating	25.2	12.1
- Appliances	15.1	2.5
- Refrigerator	3.1	−29.6
- Range	2.4	18.4
- Clothes Dryer	2.4	1.7
- Other[2]	5.6	51.4
- Lighting	5.1	15.7
- Space Cooling	1.8	165.0
Total Canadian Passenger Transportation Energy Use[3]	1.3×10^{15}	13.0
Passenger Transportation - One Small Car[4]	33.9	6.5
- One Large Car[4]	42.6	−12.0
- One Light Truck[4]	51.5	48.8

Note: 1 tonne of wood waste equals 18.0 GJ (Statistics Canada 2002).
[1]Adapted from: Natural Resources Canada 2004.
[2]"Other" includes televisions, video cassette recorders, DVD players, radios, computers, toasters, etc.
[3]Includes all forms of passenger transportation including motorcycles, school buses, urban transit, inter-city buses, air and rail.
[4]Average distance travelled per year in km: small car (16,846), large car (16,715), light truck (16,819).

The potential for bioenergy use is greatest in regions where wind, solar and hydrologic potential are low. In 1997, biomass provided approximately 5% of Canada's primary energy needs, most of which originated from the forest (Canadian Council of Forest Ministers 1997). The forest sector itself uses the largest share of the bioenergy and has increased its use by 51% from 1980 to 1997 (Canadian Council of Forest Ministers 2000), as demonstrated in Figure 1-2.

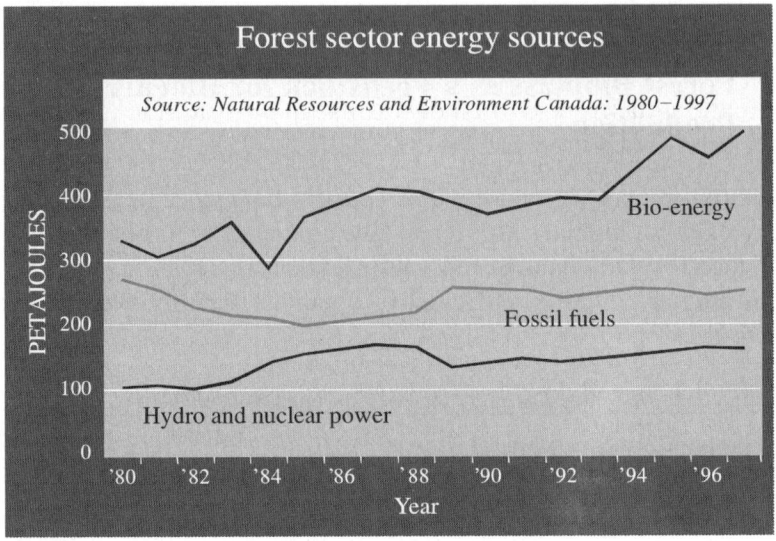

Figure 1-2. Forest Sector Energy Sources (Canadian Council of Forest Ministers 2000).

According to the Intergovernmental Panel on Climate Change (Canadian Council of Forest Ministers 2000), burning biofuels such as wood residues and pulping liquor does not result in net CO_2 emissions. Sustainable forest management includes the concept that biofuel CO_2 emissions are balanced by carbon removal, or sequestration, associated with forest growth.

From the 81.5Mt of woody forest biomass quantified in Table 1-1, it would be possible to generate 26.9 Mt or 34 billion litres of methanol fuel annually (Appendix A). This biofuel has the potential to supplement Canada's vehicle fuel consumption needs, considering that domestic sales in 2001 amounted to 38.7 billion L of motor gasoline and 22.6 billion L of diesel fuel oil (Statistics Canada 2002). This recognized biofuel market is estimated at $30.68 billion in methanol commodities (Table 1-1), not to mention jobs created laterally, environmental benefits, energy security and a decrease in health risks due to air pollution.

One of the challenges is that, compared to non-renewable fossil fuels such as oil and natural gas, wood and agricultural sources of biomass have a low energy density, leading to high transportation and handling costs. Woody material is more expensive to collect and transport than grains such as corn, barley and wheat. Woody biomass is also more costly to process due to an additional step required to breakdown cellulose into simple sugars that can be used as a food source by microorganisms for fermentation into ethanol. Once biomass is converted to fuel, however, it is easier to store and transport, and can be used for much more than just electricity generation, a limitation of solar and wind energy. As well, efforts to streamline handling and transportation of biomass resources could reduce costs and lead to an increase in the attractiveness of bioenergy use in the near future.

2.3   Composition of Woody Biomass

Woody, or ligno-cellulosic, biomass varies in composition depending on its origin, with higher lignin content in softwoods than hardwoods. Generally, this biomass is composed of 40% to 50% cellulose, 25% to 35% hemicellulose and 15% to 25% lignin (Wyman and Goodman 1993). Cellulose and hemicellulose are, respectively, six- and five-carbon chains that can be economically converted into reducible sugars and further converted into ethanol and other chemicals.

One of the current technological hurdles is the separation of lignin from cellulose and hemicellulose (Agriculture and Agri-Food Canada 2002). Lignin can be used as a high-energy combustible for boilers to power machinery or generate electricity, and also has potential as a feedstock for chemical synthesis of products, such as phenols, aromatics and olefins (Wyman and Goodman 1993).

2.4   Woody Biomass from the Forest

It is not reasonable to envision harvesting traditional timberlands to support the bioproduct industry, as it would have a two-fold negative economic impact. First, it would compete with traditional forestry employ-ment, and second, the cost of the raw material would limit the economic feasibility of such operations. Alternatively, using non-merchantable wood from current forest operations offers a large quantity of biomass at a relatively low cost.

2.4.1   Estimating Woody Biomass Availability from the Forest

The Canadian forest inventory indicates that there are 418 million hectares (Mha) of forestland, of which 227 Mha are stocked timber productive forest with a species distribution of 62% softwood, 22% mixed wood and 16% hardwood (Lowe et al. 1996). The timber productive forests currently under management consist of 120 Mha; the remaining 107 Mha consists of trees that are young (46%), mature or over-mature (42%), of uneven age or unclassified (Lowe et al. 1996). Because the bioproducts industry does not discriminate against wood characteristics such as age, size, defects and species, timber productive forests can offer great opportunities for otherwise "non-economic" resources for the industry. Thus, it should be feasible to harvest the renewable timber from these lands, utilizing the entire 227 Mha.

The average growth for Canadian timber productive forests is 360 Mm^3/year (Lowe et al. 1996). In the 120 Mha that is currently suitable and available for wood production, the annual growth rate is calculated to be 197 Mm^3 (Lowe et al. 1996). Therefore, 163 Mm^3 of wood per year is available in addition to the present harvest. To calculate the existing biomass, the volume-to-biomass conversion ratio needs to be determined, thus allowing the conversion of 163 Mm^3 into dry tonnage. Penner et al. (1997) describe precise calculations for determining biomass from specific species, age groups, stands and regions, and account for variation in height and diameter at breast height (DBH). Extrapolating from Penner et al. (1997) by using the conversion rate of 0.5 t/m^3, which represents the dry weight of all softwood, hardwood and mixed-wood tree species, the 163 Mm^3 of available wood is equivalent to 81.5 Mt of biomass available annually on a sustainable basis for the bioproducts industry (Table 1-1). Biofuel derived from this biomass has the potential to contribute over $30 billion annually to the Canadian economy (Table 1-1), as well as provide many social and environmental benefits.

2.5   Biomass from Logging Residues

Logging residues resulting from timber harvest include leaves, branches, bark, damaged and decaying logs, and intact non-preferred trees species (*i.e.*, those not valued for their timber or pulp properties). This waste biomass is often piled for slash and burned or left to decay, causing emissions of greenhouse gases into the atmosphere.

In addition to bioproduct profitability, the removal of mechanical thinnings and slash that pose a forest fire threat could reduce the need for controlled burning, which can be dangerous and costly. Collecting, chipping and extracting is expensive, however, so to become a reality, this source of biomass will need to be subsidized by parties interested in fire prevention and integrated into current forestry practices.

2.5.1   Estimating Biomass from Logging Residues

Calculating logging residues is complex due to large variations in species composition, maturity, age and size within and among ecozones. To determine the biomass available from logging operations, it is necessary to estimate the total waste produced, which will also vary with silvicultural practices. For instance, trees cut for veneer and high quality lumber will yield higher amounts of waste biomass than those cut for pulp and particleboard manufacturers. It is also difficult to gauge residues generated from damaged and decaying roundwood or from "weed tree" species.

In 1999, forest harvesting took place over an area of 1,025,429 ha and the net merchantable volume of roundwood extracted was 193.9 Mm^3, composed of the following (Natural Resources Canada 2003a):

logs and bolts	161,859,000 m^3
pulpwood	25,980,000 m^3
fuelwood and firewood	2,902,000 m^3
other industrial roundwood	3,149,000 m^3

As demonstrated in the two methods of calculation shown below, it is estimated that for every cubic meter of roundwood harvested, approximately 0.782-0.855 m^3 are left behind as logging residues.

To calculate residues as a fraction of pulpwood, Penner et al. (1997) converted inventoried biomass data into corresponding biomass component estimates. Using this methodology, it was determined that 40-year-old black spruce (*Picea mariana*) in Newfoundland consisted of 14.4% bark, 25.2% branches and 38.6% foliage, including twigs. Thus, for every debarked tonne of black spruce harvested, roughly 0.782 t of waste biomass is generated. Using this fraction, a rough estimate of logging residues could be calculated as 78.2% of all merchantable roundwood produced, which equals 75.82 million dry tonnes (193.9 Mm^3 × 0.782 × 0.5 t/m^3).

A study on coarse woody debris (CWD) in northwestern Ontario demonstrated that approximately 112 m^3/ha of downed material could be collected from mixed-wood clear cuts (Pedlar et al. 2002). By comparing stock roundwood estimates of 131 m^3/ha in the productive forest (Lowe et al. 1996) to volume of CWD (112 m^3/ha), a ratio of 1:0.855 is achieved, suggesting that for every tonne of merchantable roundwood produced in Canada, there are 0.855 t of logging residues. Using this ratio, annual roundwood production would generate 82.89 M dry tonnes of logging residues (193.9 Mm^3 × 0.855 × 0.5 t/m^3).

Conservatively speaking, the 75.82 Mt of biomass residues generated by the logging industry has the potential, via biofuel conversion, to add $28.54 billion annually to the Canadian economy (Table 1-1).

2.5.2   Potential for Increased Biomass from Logging Residues

Interior cedar hemlock (*Tsuga* spp.) ecosystems in British Columbia have a higher incidence of on-site wastes than boreal and some temperate ecosystems due to the frequency of butt and pocket rot, and weak heartwood and sapwood that increase the possibility of breakage when trees are felled and skidded. The combined decay, waste and breakage estimates range between 55% to 68% for the harvesting of western red cedar (*Thuja plicata*) western hemlock (*Tsuga heterophylla*), hybrid white spruce (*Picea glauca x engelmannii*) and subalpine fir (*Abies lasiocarpa*), depending on the harvesting techniques (Renzie and Han 2002). In this logging operation, when roundwood wastes are added to slash (crown material and branches), the roundwood-to-residue ratio could be in the range of 1:2.5, suggesting that for every tonne of merchantable roundwood produced, roughly 2.5 t remain as logging residues. This is substantially higher than the figure used to quantify biomass Canada-wide.

Also of note, the estimated biomass does not include pre-commercial thinning, which is a major feedstock for biofuel production in countries such as Finland (Hakkila 2000).

2.5.3   Environmental Concerns

One concern with the extraction of logging residues is determining what percentage of logging residues can be removed before biological sustainability is compromised. Soil compaction and removal of organic material is often purported to have negative impacts on long-term productivity in forest ecosystems (Rob Flemming, Canadian Forest Service, Great Lakes Forestry Centre, per. comm.). Hyvonen et al. (2000) found that site-remaining logging residues, 16 years after clear-felling, increased the nitrogen levels by 30% in spruce (*Picea* spp.) sites and 70% to 80% in pine (*Pinus* spp.) sites. Nitrogen concentration and release is highest in needles and for this reason, logging residues are now left on site for the needles to fall off prior to chipping and bailing.

Logging residues can also be used for the application of a thin layer of CWD, which acts in a similar fashion to that of a mulch, providing enhanced retention of soil moisture by directly binding water and refracting solar radiation. As well, CWD stabilizes logging roads and skid trails, reducing the risk of stream siltation and scarring of landscapes (Maynard and Hill 1992). The debate for biomass removal vs. retention is compounded by the fact that CWD is characteristic of old growth forests, adding to structural and inferred species diversity and composition (Brakenhielm and Liu 1998). Some authors argue the most efficient short-term management tactic to retain old-growth characteristics is to leave behind large living trees, snags and logs (Siitonen et al. 2000). These and other concerns need to be addressed when determining the quantity of logging residues to remove from any given site.

2.5.4 Examples from Other Countries

Finland has made headway with the procurement of logging residues. Similarly in Norway, government support and expenditures into green energy production, coupled with stringent goals to decrease CO_2, have led to the development of innovative equipment specialized in residue extraction. One machine, equipped with a flexible hydraulic arm in the front and bailing apparatus in the back, can make 350-400 kg bails in minutes out of disorganized, meshed slash (Dale et al. 1998). On-site chipping and bulk transportation hold promise for residue extraction with the added benefits of collecting smaller fragments and speeding up the drying process (McCallum 1999; Hakkila 2000).

2.6 Biomass from Peat

It is estimated that there are 3 trillion m^3 (Daigle and Gautreau-Daigle 2001) or 510 billion tonnes (Jasinski 2000) of peat deposits in Canada, representing approximately 25% of the world's peat resources. Nevertheless, Canada is responsible for only 5% of world production (Jasinski 2000), with less than 0.02% (17,000 ha) of its 113 Mha of peatlands being used for peat or peat moss harvesting (Daigle and Gautreau-Daigle 2001). In 1999, Canada produced 1.2 Mt of peat, representing less than 2% of the 70 Mt or more of peat that accumulates annually (Daigle and Gautreau-Daigle 2001). A large portion (800,000t) of the amount produced was exported to the United States (Jasinski 2000). Total Canadian revenues for horticultural peat in 1999 were approximately $170 M and the industry provided employment for thousands of residents in the rural areas of the nation (Daigle and Gautreau-Daigle 2001). At present, no peat in Canada is used for fuel purposes, but its carbon sequestration potential is being studied.

2.6.1   Estimated Availability and Uses of Biomass from Peat

The annual sustainable amount of peat biomass available is estimated to be 70 Mt (Table 1-1), as this is the amount of peat that accumulates annually in Canada (Daigle and Gautreau-Daigle 2001). The bulk of peat produced worldwide is used to generate energy, with the remainder sold as peat moss or used for horticultural and landscape purposes. Currently, the majority of the peat (moss) produced in Canada is sold in compressed bales for use in the horticultural and nursery industries, or for household use (Daigle and Gautreau-Daigle 2001). With emerging technologies that improve biomass-to-fuel conversion, peat biomass could generate up to $26.35 billion annually in methanol (Table 1-1). However, production will depend on developing cellulosic conversion technologies becoming cost competitive (Natural Resources Canada 2003b). Examples of existing technology that converts peat to briquettes may be used as an alternative to heating oil, and would have a value of over $18 billion.

Peat also has applications as an industrial absorbent to clean up chemical and oil spills, and can be used residentially, commercially or industrially to encapsulate fuels, blood, solvents, pesticides and other liquids (Spillsorb 2003). Peat has been successfully used to increase nitrogen (NH_4) levels in organic blueberry production (Adam et al. 2000). As well, peat can be enhanced through liquefaction and gasification processes that produce a number of by-products, such as synthetic natural gas, methane, synthetic liquid fuels and bitumen, as well as base products for chemical processing, such as phenols, tars and waxes (Coyes & Associates International 2003).

2.6.2   Environmental Concerns Surrounding Peat

Peat extraction requires that the land be cleared and the peat bogs drained, which raises ecological concerns. Extraction for horticultural peat requires that the peat remain intact and is therefore subjected to natural air-drying and delicate mechanical removal. Peat for energy consumption can be extracted by heavy equipment and subjected to mechanical or thermo-physical dewatering (Coyes & Associates International 2003).

Other environmental concerns include emissions and air pollutants generated during peat harvesting and combustion. Finland, with the highest peat production of 7.5 Mt in 1999 (Jasinski 2000), estimates that 75% of GHG emissions originate from combustion of fossil fuels and peat (Prime Minister's Office (Finland) 2000). Pollutants and CO_2 emissions generated by peat combustion obviously must be weighed against bioenergy output.

It is apparent that current levels of Canadian peat moss harvesting are not contributing to a decline in peatland functions or values. Despite the substantial room for growth of the industry, cooperative efforts between various levels of government, conservation groups and the peat industry will be required to ensure a balance between the needs of the environment and sustainable development (Daigle and Gautreau-Daigle 2001).

The appeal of peat is apparent since the bulk of Canada's peat resource is located in northern regions that are often "energy islands". As such, peat can become a critical factor in stimulating regional economic development by providing an affordable source of energy.

2.7   Biomass from Short Rotation Plantations

At its 1999 Annual Meeting, the Canadian Council of Forest Ministers identified the need to make better use of fast growing, high-yield plantations as one of the key initiatives to help Canada meet increasing global demand for wood products (Canadian Council of Forest Ministers 2001). Forest plantations can provide biomass for energy at a relatively low cost, while the establishment of these plantations does not compete with the industrial needs of forestry operations.

In Ontario, Natural Resources Canada launched the Forest 2020 plantation demonstration and assessment initiative to create new plantations (afforestation) of fast-growing tree species to mitigate GHG emissions and provide new sources of fiber (Dominy 2004). This initiative has established that the qualifying growth and yield increments be greater than $10 \, \mathrm{m}^3/\mathrm{ha/yr}$ over a 30 year, or shorter, rotation. A variety of Canadian forest tree species potentially meet this growth criteria including red pine (*Pinus resinosa*), silver maple (*Acer saccharinum*), green ash (*Fraxinus pennsylvanica*), Manitoba maple (*Acer negundo*) and white spruce (*Picea glauca (Moench)Voss*).

Red pine does well in plantations, has uniform and relatively fast growth along with good disease resistance. Depending on site quality for red pine plantations, the mean annual increment ranges from 7-15 $\mathrm{m}^3/\mathrm{ha/yr}$ (Beckwith et al. 1983). This annual growth rate is 4-8 times greater than the average yield for Canadian forests (Dominy 2004), making red pine a prime candidate for both fiber production and carbon sequestration.

Conventional fast-growing species, including hybrid poplar (*Populus* spp.), willow (*Salix* spp.) and switch grass (*Panicum* spp.), currently appear to hold the greatest potential as feedstocks for the biofuel industry. Nonetheless, other species need investigation as potential candidates for biomass farming in northern, less-productive regions such as in the Canadian taiga/tundra. Promising species such as Manitoba maple, Arctic willow (*Salix arctica Pall.*), Arctic blue willow (*Salix purpurea*), and redosier dogwood (*Cornus sericea*), will tolerate and even thrive in harsh environments and will regenerate once cropped.

According to the Poplar Council of Canada (van Oosten 2000), the *Populus* resource can be considered the "last frontier" for the forest products industry, resulting in new economic stimulus for many rural communities. Recently, there has been increased emphasis on developing management techniques leading to the successful establishment of plantations of hybrid poplar and willow species. Currently about 7,000 ha of managed hybrid poplar plantations exist in Canada, most of which have been established by the forest products industry (van Oosten 2000).

Much research has been conducted in Canada in poplar (*Populus* spp.) selection and breeding that has yielded many poplar varieties used across the country. The technology to establish short rotation plantations has advanced rapidly in the last 15 years, but further improvements in plant breeding or genetic modification may improve productivity and reduce risks of stand failure, along with disease and insect problems. While yields of fiber can be high, associated costs can be higher than traditional silviculture because of the agricultural approach to cultivation. One of the constraints to increasing yield is attributed to the high water demands of these species, as productivity is highly sensitive to water availability (Hansen 1992).

2.7.1   Provincial Policies Regarding Cultivation of Hybrid Species

Highlights of the province-specific policies affecting the future direction of hybrid cultivation are described by The Poplar Council of Canada (van Oosten 2000), as follows:

British Columbia: The provincial government classifies willow and poplar species that are intensively cultivated in plantations as primary agricultural production. The advantage of this classification is the flexibility of managing the crop with prescriptions such as pesticide application.

Alberta: The Woodlot Owners Association in Alberta has been involved in formulating a new tax policy that would qualify privately-held, managed woodlots as agricultural land for taxation purposes. Hybrid poplar cannot be planted on Crown land due to the concern over the impact this may have on the natural poplar gene pool. The proposed policy intends to include privately-held land leased to corporations, thus creating an incentive for the landowner to bring land under management while providing corporations with the means to grow intensively managed poplar crops.

Quebec: The planting of fast-growing poplar on farmland is encouraged via various subsidies available to private landowners. However, the use of all chemical pesticides disqualifies them from subsidies, thus limiting the options for weed control.

Ontario: Currently, intensively managed poplar plantations are considered a forest management activity. In 1995, the Ontario Ministry of Natural Resources ended its co-operative agreement with Domtar Papers to intensively manage hybrid poplar plantations on Domtar private and leased properties in Eastern Ontario, so the company assumed control of its own management responsibilities on these plantations.

2.7.2   Energy Balance in Biofuel Crops

When crops are cultivated as feedstock for biofuel, it is important to determine the energy balance (input/output), which is the energy generated from each crop relative to the energy input. Fertilizers generally make up 52% to 67% of energy inputs and therefore, to make energy cropping as economical and environmentally sound as possible, research should focus on species that require the least amount of fertilization.

Depending on the type of crop and technology used, each gigajoule (GJ) of energy input can provide an output of 7-50 GJ of thermally usable energy without any additional CO_2 emission (Scholz et al. 1998). This positive energy balance of 1:7-50 is astonishingly higher than that of diesel fuel which has an energy balance of 1:0.83 (Association of Equipment Manufacturers 2001), meaning that the labours involved in extracting, shipping and refining diesel fuel are greater than the energy yielded.

Field studies of different solid biofuels (Scholz et al. 1998) showed that poplar had the highest energy gain per unit area and time of 155-167 GJ/ha/yr. Rye and miscanthus had a yield and energy gain similar to poplar, but required higher energy inputs due to fertilizer and pesticide requirements, thus reducing the overall energy balance. Pine chips required the lowest energy input, attributed to the long harvest cycle (up to 140 years), but also had a low energy output or gain per unit area and time (24-27 GJ/ha/yr). Even so, the most favorable input/output ratio was achieved for pine production at approximately 1:50. Similarly favorable input/output ratios can be achieved in poplar production, provided it can be sustained over a long period without the application of nitrogen (Scholz et al. 1998).

Energy inputs such as fertilization need to be carefully scrutinized when choosing bio-fuel crop species. Life cycle assessment (LCA) methodology should be used during the planning stage when short rotation woody crops are being considered as a renewable energy option. LCA is defined by the Society of Environmental Toxicology and Chemistry as an objective process to evaluate the environmental burdens associated with a product or activity by identifying energy and materials used and wastes released to the environment, and to evaluate and implement opportunities to affect environmental improvements.

LCA takes into account the entire life cycle of the product or function and evaluates the environmental impact of the products from cradle to grave. According to Rafaschieri et al. (1999), who conducted a LCA study of poplar energy crops, the most negative environmental effect is caused by the use of chemicals and fertilizers. Therefore the ratio of biomass yield to fertilizers applied should be in the forefront of all biofuel crop planning.

2.7.3 Estimating Potential Biomass Yields from Short Rotation Plantations

Poplar is typically grown on a 10-15 year rotation at densities of 1,100-1,400 stems/ha, while willow is grown at 12,000-15,000 cuttings/ha and is harvested on a 3-4 year cycle (Samson et al. 1999). Table 1-3 shows the estimated annual yields in oven-dried tonnes (odt) of poplar and willow for regions of Canada.

Variable cash costs and full economic costs associated with production of willow and poplar are represented in Table 1-4. Variable cash costs represent the minimum market price required by producers to cover expenses related to cuttings, chemicals, fertilizers, fuel, repairs, custom harvest and transport of chips to a mill. Full economic costs also include fixed cash costs (*i.e.*, taxes) and the cost of owned resources (*i.e.*, producer's own labor and equipment depreciation). As would be expected, regions with the longest growing season and the most favorable climate have the lowest cost of production. The costs for growing willow are lower too, due to its higher productivity.

Short rotation plantations could be developed on soils that do not suit agricultural crops, thus diversifying farming activities and stimulating the rural economy. As much as 6.26 Mha (62,607 km^2) of fallow farmland are available in Canada, with nearly all (98.9%) located in the prairie provinces (Statistics Canada 1997). Based on an average poplar/willow biomass yield of 3.5 odt (oven dry tonnes) /ha/year in the prairie provinces (Table 1-3), it could be possible to harvest 21.91 Mt of biomass from fallow farmland.

Table 1-3. Estimated Yields[1] of Willow and Poplar in Canada.

Region	Poplar Yield (odt/ha/yr)	Willow Yield (odt/ha/yr)
British Columbia (lower mainland)	9-12	10-15
Prairie Provinces	1-5	2-6
Ontario and Quebec	2.5-7	7-12
Atlantic Provinces	2-6	5-10

[1]*Source*: Samson et al. 1999.

Table 1-4. Costs[1] of Growing Fast Rotation Poplar and Willow.

Region	Poplar		Willow	
	Variable Costs ($/odt)	Full Economic Costs ($/odt)	Variable Costs ($/odt)	Full Economic Costs ($/odt)
British Columbia	62-73	80-97	50-64	60-78
Prairies	104-120	146-173	92-106	115-134
Ontario and Quebec	82-112	112-160	57-82	69-102
Atlantic	91-115	127-164	64-92	78-116

Source: Samson et al. 1999.
[1]Costs were estimated using BIOCOST-Canada and BIOCOST-US software.

2.7.4   Bioenergy from Short Rotation Plantation Biomass

In the previous section, we estimate a harvest of enough woodchips from short rotation plantations (21.91 Mt @ 18,000 TJ/Mt = 394,380 TJ [Terajoule = 1×10^{12} joules]) to displace roughly 21% of Canadian coal production (69,163.1 kt in 2000 @ 27.7 TJ/kt = 1,915,818 TJ) or supplement coal consumption (1,941.9 kt in 2000 @ 27.7 TJ/kt = 53,791 TJ) 7 times over (Statistics Canada 2002). This would be an enormous step toward reducing greenhouse gas (GHG) emissions and adhering to the Kyoto Protocol (see Chapter 12), since coal is the most carbon-rich fossil fuel and produces the most GHGs when burned. Furthermore, with the adoption of the Kyoto Protocol, along with SO_2 and NO_2 permits and CO_2 taxes, the price of coal is anticipated to rise between 153% to 800% (Energy Information Administration 1998). Continuing price increases of oil and gas will also make energy crops increasingly feasible as a renewable resource to supplement all these fossil fuels.

2.7.5   Examples from Other Countries

A prototype hybrid poplar biofuel plantation in southern Wisconsin estimates yields of 6.7-11.2 t/ha of stem-dry biomass for 2-year-old regrowth (Stanosz et al. 1998). The Electric Power Research Institute has estimated that biomass production, particularly willow, could establish a new agricultural market worth as much as US $12 billion a year in the US farm-sector economy (Moore 1996).

Willow plantations in Sweden began as a result of low grain values, subsidies for energy growers, heightened taxes on fossil fuel consumption and the subsequent emergence of a biofuel market based on forest fuels

(Rosenqvist et al. 2000). According to the European Commission (1997), the European (EU15) bioenergy share of the total energy production can potentially grow from 3% to 8.5% by 2010, with half of this increase anticipated to come from energy crops.

ARBRE (ARable Biomass Renewable Energy) is the first of a program of high efficiency biomass energy generation plants to be built throughout Europe (Pitcher 2000). This project involves a 10 Mw (1 x 10^6 watts) electricity generation plant that uses willow coppice, a wood-based purpose-grown short rotation biomass, in conjunction with forestry residues to produce electricity. The plant requires 43,000 odt/yr of fuel and provides enough electricity to meet the needs of 33,500 people (Pitcher 2000).

2.8   Biomass from Sawmill Residues

Sawmill residues are composed primarily of bark, sawdust, shavings and coarse residues such as decayed, knotted, warped or insect-damaged wood. Although deemed a by-product, sawmill residues are not always considered as waste owing to secondary industries that already turn this material into products such as particleboard, electricity, wood pellets and biofuels. From a yearly harvest of 184 Mm^3 (19.93 Mt), 76% of sawmill wood residues are currently being marketed for such things as fuel production, sawdust for pulp, panels and animal bedding, forestland applications, landscaping and consumer products, (Hanam Canada Corporation 2004). As shown in Table 1-5, fuel sales amount to 40% of the total residue supply, with the non-energy demand utilizing 36%.

2.8.1   Availability and Uses of Biomass from Sawmill Residues

The total annual available biomass from sawmill residues is estimated to be 19.93 Mt based on the supply shown in Table 1-5. Presently, the most profitable products sold by sawmills are sawdust for pulping and particleboard manufacturing, and shavings for medium density fiberboard manufacturing. The current price of sawdust and shavings is $10-20 per delivered oven dried tonne (Hanam Canada Corporation 2004). A potential market for sawdust could emerge, considering that a tonne of sawmill residues could generate $367.43/t (B:C ratio of 0.33 x $1140.70) via methanol biofuel production (Table 1-1 Appendix A). If the estimated 19.93 Mt of residues were converted to biofuels, it is conceivable to create a bio-based market worth $7.5 billion.

Table 1-5. Canadian Market for Sawmill Wood Residues.

Market	odt/year '000	%
Fuel Sales	8,053	40
Sawdust for Pulp	1,807	9
Sawdust and Shavings Board	2,550	13
Sawdust and Shavings Beds	1,654	8
Other Agricultural Uses	241	1
Forest Land Applications	203	1
Landscaping	546	3
Consumer Products	6	0
TOTAL	15,060	76
SUPPLY	19,927	100
SURPLUS	4,868	24

Source: Hanam Canada Corporation 2004.

Wood residue power generation is proving viable in projects across the country. This is demonstrated at Domtar's White River, Ontario sawmill operation that has formed a partnership with Primary Power International. The White River BioMeg plant was commissioned to consume a 25-year-old bark pile adjacent to the Domtar mill (Primary Power International 2005). This 7.5-MW plant consumes 50 T of waste wood feeds and fuel from the stockpile each day. A portion of the steam generated is sold back to Domtar and the rest is sold through a power purchase agreement (PPA) to Ontario Hydro to be sold on the provincial grid. The 7.5 MW of power is enough to operate the Domtar mill, the town of White River and the plant itself, thus making the Domtar mill 100% energy efficient (Mohammed 1999). Through utilization of biomass on site, transportation costs of the raw material are avoided. This is advantageous since transportation typically consumes most of the wood residue price, resulting in a minimal return of less than $1 per dry tonne to the sawmill (Hanam Canada Corporation 2004).

2.9   Municipal Waste Wood and Paper

Canada is second only to the US in its use of paper and paper products. As such, paper wastes are one example of a landfill product with enormous consumer products and bioenergy potential. Nearly 5 Mt of used paper products are disposed of annually in Canada, representing 35% (by weight) of solid wastes in municipal landfills (Canadian Council of Forest Ministers 1997). Disposal of these products in landfill sites results in carbon being returned to the atmosphere through decomposition.

According to the David Suzuki Foundation (2004), methane emissions from Canadian landfills produce the same amount of GHG as 5 million cars, and are expected to increase by 19% by 2010.

Through recycling and the conversion of woody municipal solid waste (MSW) into commodities, decomposition is halted, and GHG emissions into the atmosphere are reduced. Recycling of paper products thus plays an important role in the forest carbon budget by reducing the amount of material ending up in landfill sites and, ultimately, the requirements for harvesting (Canadian Council of Forest Ministers 1997).

2.9.1   Estimating Available Biomass from Municipal Waste Wood and Paper

To obtain a better understanding of the relationships between municipal solid wastes, recycling and bioproducts, we need to recognize the existing situation of MSW and recycling in Canada, as well as the availability, characteristics and potential of waste wood and paper.

It is difficult to accurately estimate the amount of wood and paper material in MSW, in part due to changes in survey methodology used by Statistics Canada over the years (Sherry Vermette, Statistics Canada, per. comm.). In 1992, waste materials were segregated by type (Figure 1-3), whereas recent surveys segregate waste by source only (Table 1-6).

In 2000, 31.4 Mt of non-hazardous MSW material were generated in Canada, corresponding to 746 kg of waste for each Canadian, up 9% from 1998 (Statistics Canada 2003). The majority of this waste (51%) originated from the industrial, commercial and institutional (IC&I) sector, while 36% came from residential sources, and 12% were from construction and demolition (C&D) sources (Statistics Canada 2003). Of this total, 23 Mt were disposed of at municipal and private waste facilities, 0.9 Mt was exported and 7.5 Mt was processed for recycling (Table 1-6). The total amount of wood and paper products recycled in 2000 was approximately 3.44 Mt (Statistics Canada 2003).

Through application of the data represented in Figure 1-3 and Table 1-6, the annual biomass available from waste wood and paper is estimated to be 8.036 Mt (31.379 Mt total waste generated x 25.61%, Table 1-1). This total includes 4.596 Mt of waste wood and paper within MSW facilities in addition to 3.440 Mt that were identified as recycled material in the year 2000 (Statistics Canada 2003). Of this total, an estimated 3.138 Mt (10%) is waste wood and 4.898 Mt (15.61%) is waste paper and paperboard.

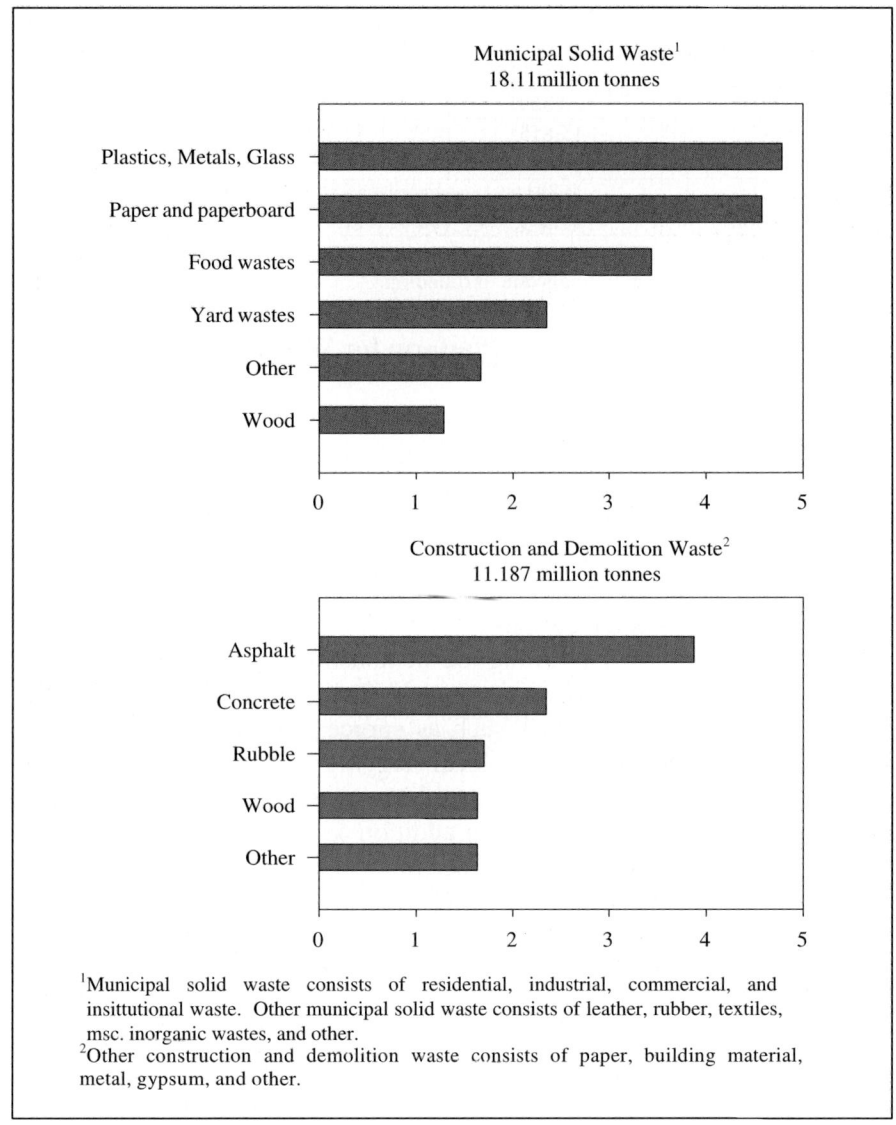

Figure 1-3. Materials in the Solid Waste Stream, 1992 (Adapted from: Senes Consultants Ltd. and Frankin Associates Ltd. (1995) in: Environment Canada 2003).

Municipal Waste Tonnage (1992)

Total (all sources, including recycling)	29.297 Mt
Paper and paperboard	4.576 Mt (15.61%)
Wood	2.919 Mt (10.00%)
Total wood and paper (excluding "other" MSW)	7.495 Mt (25.61%)

Table 1-6. Disposal of Waste by Source in Canada, 2000.

Waste Source	Total Waste by Source in Mt	MSW Disposal by Source in Mt	Materials Recycled by Source in Mt
Residential Sources	10.869	8.350	2.0519
IC&I Sources	15.815	11.799	4.016
C&D Sources	3.802	2.836	0.977
Net Exports	0.893		
Total Generation	31.379	22.985	7.502

Source: Statistics Canada 2003.

Note: Figures may not add up to totals due to rounding.

2.9.2   Establishing a Trading Platform for Waste Wood Material

A market is certainly emerging for recycled wood materials, especially demolition waste. Fifty material recycling facilities have opened in major Canadian urban markets since 1988 (Canadian Council of Forest Ministers 1997). These facilities sell wood products at 30% to 50% below the cost of new materials and have spawned technologies to manufacture new products such as animal bedding, landscaping mulch, car floor mats and door panels from these recycled wood materials.

These alternatives to landfilling are demonstrated by "And Sons Waste Knot", a broker of waste wood material in Sault Ste Marie, Ontario. This company collects wood residues such as spruce, poplar, balsam fir (*Abies balsamea*), pine and bark from local logging and sawmill operations. Limited to a 350 km radius due to cost of transport, the company transported 85,000 tonnes of residues in 2002, in addition to loads delivered directly to buyers. The materials, which sometimes include grade B particleboards and some municipal arboricultural waste, are mechanically chipped/mulched and then sold to manufacturers who generate steam for medium density fiberboard and chipboard production, to pulp and paper mills as hogfuel (bark), and to local landscapers as garden mulches.

Establishing a market place or a trading platform for buying and selling waste wood materials could provide a steady and reliable stock of resources, ensuring that supply can meet the demands of processors. In addition, the establishment of a market will help determine the value of waste from the forest products pool while increasing profits and decreasing the need to consume other resources. A European online market, IHB Timber Exchange (2004) is a prototype, offering the wood manufacturing industry access to wood products that might otherwise end up in landfill sites.

2.9.3 Environmental Concerns about Waste Wood Material

The large piles of wood and other residue forest product biomass that collect in urban disposal sites can be a valuable source of raw material. The importance and advantage of these waste materials as a feedstock for bioenergy are due to their abundance and low, or often negative, cost. However, municipal waste wood originates from a variety of sources and most is contaminated to some extent. Contaminated waste wood includes the by-products of demolition, home renovation and industry, which usually contain nails, hinges, paints, plastics or cements. As well, manufactured products such as medium density fiberboard (MDF), flakeboard and plywood all contain adhesives. Waste wood generated from urban arbori-culture is considered wholesome until blended with municipal waste wood. If economically viable, municipal waste facilities can sort their on-site collections into categories depending on their characteristics.

It is uncertain at this time if the waste wood estimates include guard rails, utility poles and rail ties, which are often abandoned, and due to their rot-proofing (creosote), would need some special attention. As such, this waste would need to be managed more stringently to ensure that health and environmental safety are not compromised. These materials are prime candidates for industrial gasification, which eliminates harmful air pollutants through filtration and physiochemical processes.

2.9.4 Conversion of Waste Wood and Paper to Biofuel

With the establishment of the bioeconomy, the 8.04 Mt of waste wood and paper could render a value of $3.03 billion annually via conversion to methanol (Table 1-1). This value is comparable to the current operating revenues generated within the waste management industry, totaling $3.4 billion in 2000 and employing 30,980 people across Canada (Statistics Canada 2003). The establishment of waste-derived biofuels will contribute to the solution for waste management, GHG concerns and energy security.

Despite these benefits, conversion to methanol may not be the most practical use of waste paper material. Although mixed paper and organics have the best energy producing potential within MSW, recycling this material into consumer products is more typical (Wood and Layzell 2003). Interestingly, if all waste material is considered, MSW contains an average 85% combustible material from which energy may be derived from technologies including combustion, pyrolysis, gasification and biogas production through aerobic or anaerobic fermentation.

2.9.5   Conversion of Waste Paper to Levulinic Acid

Currently, levulinic acid, a high-value platform chemical, is processed from fossil fuels and averages US$1 per pound (lb) to produce and is sold for $10,000 a ton to the specialty chemical market. Conversely, in upstate New York, BioMetics Inc. (along with the US Department of Energy and the New York Energy Research and Development Authority) are constructing a 5000 ft^2 plant that will process waste paper that cannot be recycled into levulinic acid, (Fahy 1996). This innovative plant will have the capacity to produce levulinic acid from waste paper biomass for as little as five cents a pound (Fahy 1996). This describes but one more example of creative uses of technology forging new partnership opportunities within the bioeconomy.

2.10   Biomass from the Pulp and Paper Industry

Between 1991 and 2001, the Canadian pulp and paper industry consumed approximately 14 Mt of wood pulp annually (Natural Resources Canada 2003). The typical kraft pulping process produces roughly 50 kg of dry waste (high in organic matter) per tonne of wet pulp (Belsito and Winterhalder 2000). The primary waste is sediment composed of cellulose fibers (50% to 60%), fines such as clay (10% to 25%) and calcium carbonate (10% to 25%). Secondary waste sludge results from the microbial treatment of effluents in aerated lagoons or activated sludge reactors and is mostly comprised of microbial biomass (70% to 75%) along with some fine inorganic particles (25% to 30%), such as clay and calcium carbonate (Belsito and Winterhalder 2000).

Log yard residues (LYR) result when bark and fine wood particles fall from the log and mix with soil and rock. High ash content limits the usefulness of LYR as a biofuel (Campbell and Tripepi 1992), while high organic content restricts its use as fill material. Since LYR are often land-filled or stockpiled, this can cause environmental problems due to sponta-neous combustion or leaching of acidic wood extracts into groundwater. However, bark and wood residues can be suitable raw materials for fuel pellet production if the ash content is controlled, with the benefit of high durability of bark pellets compared to sawdust pellets (Lehtikangas 2001).

2.10.1   Estimating Biomass from the Pulp and Paper Industry
and Biofuel Potential

Despite variations in sludge content and mill techniques, 4245 odt of paper mill sludge are produced daily in Canada (Emery 2003). We have used this figure to estimate the annual biomass from paper mill sludge to be 1.55

Mt (Table 1-1). If all paper mill sludge biomass were converted to alcohol, a biofuel industry could be established with the potential to contribute $350-$580 M annually to the Canadian economy.

The pulp and paper industry can use pulp liquor and LYR on-site to produce steam and generate electricity. For example, Howe Sound Pulp and Paper Limited in British Columbia has reduced operating costs by supplementing its energy requirements with black liquor and hog fuel (bark). To achieve this, the mill has constructed three bark presses, which produce pellets and briquettes for fuel, resulting in elimination of bunker C oil consumption (a term used to describe the most thick and sticky of the residual fuels) along with a reduction in natural gas use. Immediate environmental benefits for this company include a reduction of CO_2 GHG emissions equaling 8.3% over the past 10 years (Howe Sound Pulp and Paper Ltd. 2000).

Major barriers to realizing benefits from pulping liquor include dewatering and the costs associated with transportation. Often the primary and secondary wastes are recombined, and dewatered to a 25% to 35% solid for ease of handling. However, the costs involved in dewatering are often not recovered, resulting in much of this material being landfilled instead.

2.10.2 Use of Papermill Sludge as a Land Application/Soil Amendment

At one point in time, land applications of paper mill sludge would have been unsafe due to the presence of toxic compounds such as dioxins, furans and PCBs. Replacing peroxides with chlorine has significantly lowered dioxin and furan problems, and the elimination of carbon-copy paper in the 1970s has reduced the PCB problems. Improved paper manufacturing technology has resulted in pulping liquors with lower levels of heavy metals and organic contaminants than municipal sewage sludge (Belsito and Winterhalder 2000). This makes the sludge a prime candidate for land applications, especially in disturbed or contaminated areas.

A trial land application of 20 dry t/ha of Domtar's Windsor sludge proved successful due to high nitrogen levels and a good carbon-to-nitrogen ratio. The levels of furan and dioxin were found to be well below levels recommended by the Canadian Council of Ministers of the Environment and the US Environmental Protection Agency (Sylvestre et al. 1999).

In northeastern Ontario, a study investigating the reforestation of ecologically disturbed areas examined the effectiveness of soil amendments and land coverings by comparing horticultural peat, humus from jack pine (*Pinus banksiana*) forests and natural lake sediments to that of paper mill sludge. A 10-cm layer of sludge was applied to disturbed areas, including gravel pits, old roadways, areas with marginal soils (incapable of holding water) and sites contaminated with heavy metals where vegetation was unlikely to establish. The authors concluded that paper mill sludge was the most effective of the aforementioned soil amendments, demonstrating an ability to establish vegetation of great diversity (Belsito and Winterhalder 2000).

In another study (Simard et al. 1998), paper mill sludge compost applied to potato fields significantly improved the soil (improved nutrient retention and decreased leaching) as well as potato quality (degree of potato scab was reduced) and yields. Use of the compost reduced the need for inorganic fertilizers, resulting in higher profits for the farmer, in addition to providing a waste management solution for the pulp and paper industry.

Paper mill sludge could also be applied to sulphide-tailing operations as a means of buffering acid leaching. This potential is locked within the fibrous composition of paper biomass and, if feasible, could be at the forefront in the reclamation of tailing ponds. The benefits of using organic matter in covering sulphidic mine tailings is that it provides both an oxygen barrier and oxygen consumer, thus producing fewer airborne sulphur contaminants (sulphate gas emissions). The organic matter can form complexes with metals such as copper, and thereby reduce the phytotoxic properties of the growth medium (Belsito and Winterhalder 2000). A further advantage is that typical organic coverings, such as peat and black muck, involve the use of primary resources, whereas paper sludge is a waste product and otherwise requires disposal.

3.   CONCLUSION

This chapter provides an overview of Canadian forest biomass quality, and estimates the quantity of biomass available from various sources for exploitation by the bioproducts industry.

There is clearly a wide variety of options through which forest biomass may be used in an environmentally responsible manner for the production of various bioproducts commodities. At present, the majority of this biomass tends to be either neglected, as in the case of logging operations, or underutilized, as in the case of peat and municipal waste wood and paper. In most instances, obstacles must be overcome to efficiently realize the value of these biomass resources. While some conversion options are presently feasible, others will rely on environmental and economic viability (*i.e.*, petroleum sector competition), developing future technology, availability and distribution of the biomass, along with societal acceptance.

It would be beneficial to the establishment of the bioproducts sector to fully understand the cost savings that result from waste biomass utilization and the technical approaches applicable to each bioproduct commodity. If, for example, logging residues from roundwood production could simultaneously be obtained with virgin biomass, the overall cost of infrastructure for roundwood and biomass procurement could be reduced.

Understanding of the novel potentials of the various types of biomass demands close monitoring and inventory of these residues, as well as delineation of a proper procurement plan complete with demonstration projects. Monitoring and posting waste residue inventories could allow industries to analyze and predict the overhead costs and benefits of adopting new technologies. While waste is being converted into bioproducts, economic savings and environmental benefits accrue, thereby stimulating a reduction in waste management, dumping fees, freight costs and GHG emissions.

One of the most lucrative options is the conversion of biomass to various forms of consumable energy, although many other options exist. Paper mill sludge conversion to a soil amendment is one emerging alternative for waste that could otherwise be destined for landfill.

With the proper incentives, implementation of a bio-based economy can proceed rapidly. Past successes demonstrate that the production of bio-energy and bioproducts, including the more traditional non-timber forest products, can be economically feasible and, in many instances, stimulate more profitable rural economies.

4. APPENDIX A

4.1 Appendix A-1

Appendix A-1: Calculations used to Determine Available Biomass, B:C Conversion Ratios and Commodity Prices.

Annual Sustainable Dry Biomass by Type	Calculations
Forest Biomass 81.50 Mt	163 Mm^3 of extant wood biomass from timber productive forest (Lowe et al. 1996) with a conversion rate of 0.5 t/m^3 (Penner et al. 1997).
Logging Residues 75.82 Mt	193.9 Mm^3 of roundwood (Natural Resources Canada 2003a) at residue:roundwood ratio of 0.782 at 0.5 t/m^3 (extrapolated from Penner et al. 1997).
Peat 70.00 Mt	70 Mt of peat accumulate annually in Canada; cited in Daigle and Gautreau-Daigle (2001).
Short Rotation Plantations 21.91 Mt	Harvesting 3.5 odt/ha/yr (Samson et al. 1999) in the 6.26 Mha of Canadian fallow farmland (Statistics Canada 1997).
Sawmill Residues 19.93 Mt	The supply of sawmill wood residues in the Canadian market was cited as 19.93 Mt (Hanam Canada Corporation 2004)
Municipal Solid Waste Wood and Paper 8.04 Mt	Total waste generated, 31.379 Mt (Statistics Canada 2003); percentage of wood and paper in municipal waste, 25.61% (Environment Canada 2003).
Papermill Sludge 1.55 Mt	Canadian sludge production of 4245 dry t/day (Emery 2003); annual amount based on 365 days of production.

4.2 Appendix A-2

Appendix A-2: Commodity Prices and Potential Biomass-to-Commodity Conversion Ratios used to Determine Total Production and Gross Economic Values in Table 1-2.

Commodity	Biomass to Commodity Conversion	Commodity Price ($Cdn)
Ethanol	Based on softwood forest residues, 100 kg dry matter converts to 20 kg = 0.20 (Eklund and Pettersson 2000)	Priced to compete with gasoline at $0.85/L × 1342 L/t (Statistics Canada 2002) = $1140.70/t
	Comments: "Peat" cellulosic ethanol technologies are still being developed and are not yet cost-competitive (Natural Resources Canada 2003b).	
Methanol	1.0 t of biomass = 417.12 L of methanol (Clean Fuels Development Coalition 2002) × methanol density of 0.79 g/mL = 329.5 kg/t = 0.33 B:C ratio.	Priced to compete with gasoline at $0.85/L × 1342 L/t (Statistics Canada 2002) = $1140.70/t
	Comments: Methanol production via pyrolysis converts peat at the same ratio (0.33) as wood (Krigmont 1999).	

Commodity	Biomass to Commodity Conversion	Commodity Price ($Cdn)
Bio-oil	1 t of dry woody biomass (6% moisture) produces 75% w/w bio-oil, 13% w/w char and 12% w/w combustable gas (Freel and Graham 2000)	16.53 M BTU/t (ENSYN Group Inc. 2001) / 36592 BTU/L × $0.569/L for heating oil equivalent (TDC Marketing and Management Consultation 2004) = $257.04/t.
Hydrogen	Bio-oil produces 6.4% hydrogen (Czernik et al. 1999; ENSYN Group Inc. 2001) via fast pyrolysis/catalytic steam reforming. B:C of hydrogen = 0.048 (from woody biomass bio-oil).	1kg H_2 is energy equivalent to 3.75L gasoline × 0.85/L (National Academy of Engineering and Board on Energy and Environmental Systems 2004) = 3187.50/t
Woodchips	Assume 1:1 conversion ratio of wood into wood chips. At 45% moisture content, one ton of wood chips displace 211 L of heating oil (McCallum 1999).	Priced to compete with heating oil at $0.569/L (TDC Marketing and Management Consultation 2004) = $120.06/t (McCallum 1999).
	Comments: Assumes 80% system efficiency for oil burners and 70% system efficiency for woodchip combustors.	
Pellets / Briquettes	Assume 1:1 conversion ratio of biomass to pellets. Softwood pellets at 5-6% moisture = 19.78 M BTU/t (Energex 2004).	19.78 M BTU/t / 36592 BTU/L × $0.569/L for heating oil equivalent (TDC Marketing and Management Consultation 2004) = $307.54/t.
	Comments: Peat briquettes at 12% moisture = 16.74 M BTU/T (Bord na Móna 2001) = $260.30/T. Paper mill sludge = 3.54 M BTU/T (Albertson and Pope 1999) = $55.05/T.	
Levulinate	Conversion of cellulose to levulinic acid is 2:1 (US Department of Energy 1998). Assuming wood-derived biomass, including papermill sludge, contains 50% cellulose (Belsito and Winterhalder 2000), the B:C conversion = 0.25	Currently selling for $1487.00/T (Fahy 1996)
	Comments: "Peat" cellulosic technologies are still being developed and are not yet cost-competitive (Natural Resources Canada 2003b).	
Electricity	Wood energy = 4.2 mWh/T (Network for Alternative Technology and Technology Assessment 2000).	At $60.00/mWh × 4.2mWh/T = $252.00/T
	Comments: Sod Peat at 35% moisture = 12.11 M BTU/T (Bord na Móna 2001) / 3414 BTU / kWh (Network for Alternative Technology and Technology Assessment 2000) = 3.5 mWh/T = $212.83/T. Paper mill sludge at 3.54 M BTU/T = 1.04 mWh/T = $62.21/T.	

Chapter 2

BIOFUELS AND BIOENERGY

Highlights and Fast Facts

🌲	Biomass-derived bioenergy, together with hydropower and wind power, could make Canada's energy supply 100% renewable.
🌲	Various biofuels have the potential to add $22-105 billion to the Canadian economy.
🌲	Bioenergy from forest-derived biomass has the potential to supplement 1.4% to 39% of annual energy production and 3% to 76% of annual energy consumption in Canada.
🌲	Bioenergy generated from biomass is carbon neutral and can both contribute to environmental protection and stimulate employment.
🌲	Establishing a reliable biofuel economy will require government efforts to encourage its use.

Forest biomass, whether found in the forest ecosystem or associated with logging and mill wastes, can originate from timber productive or non-timber productive forests, managed or unmanaged stocks, or agroforestry systems. The objective of this chapter is to determine the potential contribution of forest biomass to Canada's energy sector.

There are many ways in which biomass can be converted into useful forms of energy such as heat, electricity or biofuel. Direct combustion produces heat and if the heat is captured in the form of steam, it can be used to produce electricity. Biological, thermal and chemical conversion processes can convert biomass into fuels that can be used in vehicles, such as ethanol or methanol. The gasification of biomass produces a mixture known as synthesis gas, or syngas, which can be burned for energy, refined to make hydrogen, or used to produce other chemicals and fuels (Greene 2004).

While the forest industry already obtains more than half of its energy needs from renewable biomass (Canadian Council of Forest Ministers 2000) and is sometimes entirely self-sufficient (*e.g.* Domtar's mill in White River, Ontario, See Chapter 1), it has yet to engage in the large-scale manufacturing of fuel as a commodity. With globalization of markets and an increase in demand for value-added bioproducts, there would be definite advantages to integrating the bioproducts and timber industries. At the same time, the goal of sustainable forest communities could be accomplished.

Over the last century, motivated by abundant and inexpensive sources of fossil fuels, petroleum-based industrial products have replaced products originally made from biological materials. Today, however, there are pressing reasons to return to bioenergy and bio-based products. These include:

1) New technologies: Progress in biotechnology and process engineering has increased the feasibility of transforming almost any type of biomass into biofuels, in particular the conversion of biomass to sugars and other chemicals. Although there is still a great need for research and development to reduce the cost of bio-based industrial products, the path has been laid out for western economies to progressively shift to the use of biofuels.

2) Energy security: Since almost half of Canada's energy needs are imported, there is a risk of economic and political uncertainty associated with fossil fuels as the primary source of energy (Statistics Canada 2002). It is wise for Canada to develop alternative fuels that are less susceptible to international market fluctuations and political instability.

3) Environmental concerns: Biofuels are considered carbon neutral, whereas fossil fuels are a major source of pollution and greenhouse gas (GHG) emissions. Unlike other carbon-neutral sources of energy such as solar, wind, tidal or nuclear, the use of forestry, agricultural, marine and urban wastes as feedstocks for the bioenergy market provides a multi-faceted solution to disposal of society's wastes. By using biomass residuals such as paper sludge, waste wood, urban garbage and paper products, there is also less need to log virgin forests as a source of biofuel.

4) Increased employment: The potential exists to double energy sector employment by shifting from fossil fuels to biofuels. Biofuels could stimulate prosperity in northern and rural areas of Canada where there is a wealth of biomass, yet unemployment. The harvesting of biomass (such as logging residues and waste wood) and the processing and refining of new commodities can create jobs and increase the profits returned from logging operations.

5) Economic incentives: Costs incurred from purchasing foreign crude oil can be redirected to local economies, benefiting factories or loggers. In particular, domestically-sourced biofuels, like ethanol, have the potential to create an export market and billions of dollars in investment opportunities. Governments may lose revenue through tax incentives but will gain more from personal and corporate income taxes, and see a reduction in social security and unemployment costs.

Other countries have led the way by funding research and development into "green fuels" as well as by providing tax incentives for their use. Scandinavian countries, for example, have imposed petroleum, emission and environmental taxes to decrease dependency on petro-fuels.

In 2001, bioenergy production in Canada was 590 PJ (0.164 TWh) and provided only 5% of primary energy needs (Hall 2001). Table 2-1 provides an overview of Canada's current energy production and consumption.

Table 2-1. Canada's Energy Balance: Production and Consumption of Energy for 2000.

Energy Source	Production (Tj)	Production (billion L)	Consumption (Tj)	Consumption (billion L)
Coal	1,509,905		52,265	
Diesel	947,750	24.515	903,871	23.38
Gasoline	1,460,988	42.152	1,328,795	38.338
Natural gas	7,062,109		2,353,409	
Electricity	1,524,557		1,812,245	
Total	12,505,309		6,450,585	

Source: Statistics Canada 2002.

Table 2-2 shows the potential energy value of forest biomass if it were converted to each of the various commodities (ethanol, methanol, bio-oil, hydrogen, wood chips, pellets-briquettes or levulinic acid).

Table 2-2. Potential Energy Value by Commodity, of Converted Forest Biomass.

Commodity	Amount of Biomass Available for Conversion (MT)	Biomass to Commodity Conversion Ratio	Potential Amount of Commodity (MT)	Potential Energy Value (TJ)[1]
Ethanol	208.75	.20	41.75	751,500
Methanol	278.75	.33	91.99	1,655,775
Bio-oil	207.2	.75	155.4	2,797,200
Hydrogen	207.2	.048	18.40	331,155
Wood chips	182.37	1.0	182.37	3,282,624
Pellets-Briquettes	273.85	1.0	273.85	4,929,264
Levulinic acid	208.75	.25	52.19	939,375

Source: Table 1-1.
[1] 1 MT wood = 18,0000 terajoules (Statistics Canada 2002).

Table 2-3 shows the potential of each of these commodities to replace current energy consumption, from a low of 3% for hydrogen to a high of 76% for pellets-briquettes. Not indicated in this table is the potential for bioenergy to supplement 1.4% to 39% of energy production.

Table 2-3. The Potential of Bioenergy, Expressed as a Percentage, to Replace Canada's Annual Energy Consumption, Based on Energy Value.

	Coal	Diesel	Gasoline	Natural Gas	Electricity	Total
Ethanol	1438	83	57	32	42	12
Methanol	3168	183	135	70	91	26
Bio-oil	5352	309	211	119	154	43
Hydrogen	343	19	13	8	10	3
Wood chips	6281	363	247	139	181	51
Pellets-Briquettes	9431	545	371	209	272	76
Levulinic acid	1797	104	71	40	52	15

These tables will be referred to in each subsection when discussing the potential energy value and market worth of the various commodities. By implementing newly developed technologies and ideas, waste biomass could be used to generate biofuels that would supplement and eventually replace petrochemicals.

Cost efficiencies

While forest biomass is a relatively inexpensive feedstock, especially when waste material is used, there are drawbacks. It is costly to move and to store. To obtain the highest possible yield, energy consumption by whatever conversion process must be minimized. Certainly, mobile equipment that can wholly or partially process biomass at the source would help reduce transportation costs.

Examples of sustainable fuel development in other countries that optimize input/output ratio should be considered as models for forest-derived biofuel production. In Brazil, for example, sugar from sugar cane is fermented to alcohol as a fuel source. The steam produced in the process generates electricity, which is then used to run the plant and make fertilizers, while the thrash is turned into pellets to fuel the boilers. The surplus energy produced is sold as electricity to the grid. This model practice could be applied to the use of forest biomass (bio-oil produced from plant extracts, ethanol from wood, hydrogen and methanol from wastes), bringing Canada a long way towards supplementing or replacing the petrodiesel market.

1.   ETHANOL

Canada's annual ethanol production is approximately 240 million litres, a fraction of that of the United States (6.1 billion litres in 2000, projected to increase to 28.4 billion litres by 2016, Agriculture and Agri-Food Canada 2002). The Canadian government is aiming to increase production to 1 billion litres by 2010 (Agriculture and Agri-Food Canada 2002), while the Canadian Renewable Fuels Association expects ethanol production to top 5 billion litres in 2005. The fact that various provincial governments are adopting legislation to mandate a percentage blend (5% to 10%) of ethanol in gasoline, suitable for all automobiles, certainly will encourage this trend.

Canadian markets also appear to be receptive. The recent development of *E10 gasohol* (a blend of 10% ethanol and 90% gasoline) has led to the expansion of over 1,000 fueling stations across Canada supplying the gas-alcohol blend (Canadian Renewable Fuels Association 2004). Construction of new plants, with federal support, is planned for Seaway Valley, Cornwall and Varennes, Quebec, to increase ethanol production by 750 million litres annually (Canadian Renewable Fuels Association 2004), using both corn and wheat as feedstocks.

For the forest industry, emerging technologies are enabling ethanol production from cellulosic feedstock such as waste wood, logging residues and papermill sludge. If all the available biomass were dedicated to ethanol production, 41.75 million tonnes could be produced, enough to displace 57% of current gasoline consumption on an energy-equivalent basis (Table 2-3). Priced to compete with gasoline at 85¢ per litre, it would be possible to establish a market worth close to $47 billion annually (Table 1-1).

The wholesale cost of a litre of ethanol in Canada is about $.63, or $.38 including a tax rebate of $.25 (Agriculture and Agri-Food Canada 2002). Canada's CANMET Energy Technology Centre estimates the current cost of producing a litre of ethanol from wood in Canada to be $.30-$.35 and that is predicted to drop to $.22 within a decade (CANMET Energy Technology Centre 2003). Through research and development, the US has been able to reduce the price of producing a gallon of ethanol to about US$1.00 (US$.27/L), a tremendous decrease from US$1.43 (US$.38/L) ten years ago.

CANMET Energy Technology Centre is working in partnership with industries and institutions, such as Iogen Corporation, University of British Columbia, Tembec Inc., Kernestrie Inc. and Commercial Alcohols Inc., to bring waste-derived ethanol to the marketplace. In April 2004, Iogen announced it had produced the world's first cellulose ethanol fuel (made from agricultural-based feedstock) for commercial use (Iogen 2004). Tembec's Silvichemicals Group currently is the only company to use wood feedstock. Tembec uses acid-based pulp liquor, which seems best suited for ethanol production from cellulose because the acid hydrolysis helps break down the hemicellulose prior to the fermentation process. The company produces 17 million L of ethanol (Tembec 2004) which is further fermented into dilute acetic acid, or vinegar. The vinegar is used in eastern Canada for pickling and food processing, and as an industrial solvent for such things as pharmaceuticals, cosmetics and coatings. In fact, ethanol can be further processed for production of ethylene, succinic acid and other "platform chemicals" used in manufacturing and value-added processing (further discussed in Chapter 3). Ethanol production also assists in establishing other markets via co-production (during refining and lignin extraction) of nutraceuticals, pharmaceuticals, bioplastics, biopesticides, essential oils and silvichemicals.

1.1   Research and Development on Ethanol Production

The first step in ethanol production is the pre-treatment of wood using steam or extrusion-based processes, which needs to be optimized to handle large volumes for mass production. Next, cellulosic material is pre-digested (hydrolyzed) into monomeric reducible sugars, since the yeast used in the fermentation process cannot break down the highly branched sugars found in wood cellulose. Not surprisingly, the effectiveness of cellulose hydrolysis is limited by the effectiveness of the enzymes used. The process requires mass production of enzymes in bioreactors, and ways of lowering the currently high cost of this method of enzyme production need to be found.

Overcoming the differential rates of hydrolysis for cellulose and hemicellulose needs to be further examined. Work is also needed in separating the lignin from the cellulose and hemicellulose, (not an issue for pulp residues) since the extraction and recovery of lignin is a major contributor to the overall cost of production (Gregg et al. 1998).

A techno-economic model developed by Gregg and Saddler (1996) evaluates the effect of enzyme recycling and hydrolysis time on the production cost of ethanol from both hardwood and softwood substrates. The enzyme requirement represents about 18.5% and 22.7% of the total cost for ethanol production from hardwood and softwood, respectively. Such improvements could very well help reduce the cost of ethanol production.

There are several cell-wall-degrading enzymes used in research experiments to dissolve cellulose, pectin and lignin, and currently researchers are attempting to isolate proficient enzymes from bacteria and fungi. Research is ongoing in the development of the enzyme cellulase, which breaks down cellulose. The gene that codes for cellulase has been isolated by scientists at Cornell University and is grown in large quantities by *E. coli*.

Although still in the developmental stages, these findings show how rDNA technology could contribute to efficient biofuel production (Schor 1994).

In the fermentation stage, yeast (*Saccharomyces cerevisiae*) and other microbes are required. The choice of microbes, growth conditions, and other exogenous factors such as fibre type and toxic contamination, can limit or slow the fermentation process. Work on improving filtration of various sugars that are difficult to ferment, such as pentose, is also required.

Hardwoods have lower amounts of lignin but higher pentose sugars, whereas the reverse is true for softwood. In the fermentation process, only about 47% of sugars are converted to ethanol. New research, however, has found that the application of a low voltage can enhance ethanol production by 12% and decrease acetate production (Shin et al. 2002).

Improving the efficiency and cost of ethanol production also will require enzyme recovery, ethanol recovery, and the recycling of microbes and undigested sugars. As well, it could be advantageous to develop "biomass hybrids" with a lower lignin content and increased growth rate.

1.2   Future Directions

From the industrial point of view, obstacles to ethanol production from forest biomass include the up-front expense of building a large-scale test plant and the establishment of reliable biomass suppliers to fuel the large feedstock demand of 45.4 tonnes per day (Agriculture and Agri-Food Canada 2002). To cut costs, conversion of a paper mill already equipped to distribute large volumes of material, with large casks for processing, might be considered.

If ethanol is to compete with petroleum-based products, especially while the infrastructure is becoming established and the operation scaled-up, government incentives and other economic instruments become critical. Ethanol can be made less expensive than gasoline, for instance, with tax exemptions on E10. The US government is supplying up to US$30,000 in tax incentives for fueling sites to install US E85 infrastructure. A long-term commitment will be required in both countries to realize the net economic, environmental and social benefits.

Consumer awareness is another focal point for ensuring the successful adoption of ethanol fuel. Information such as the CANMET Energy Technology Centre fact sheets, *Ethanol the 'Green Gasoline'*, could be distributed at gas stations where fuel sales are anticipated. Raising the profile of biofuels in the media and on the internet would also be helpful in engaging the public.

2.   METHANOL

Methanol can be used as a transportation fuel in its pure form, as a feedstock for the gasoline additive methyl tertiary butyl ether, or as a carburant in fuel cells. It is cleaner burning than gasoline, storable, transportable and efficient for use in gasoline engines. Producing methanol from biomass also offers significant economic, environmental and energy advantages.

Methanol can be produced from biomass using a two-step thermo-chemical process: gasification to produce hydrogen and carbon monoxide, then a reaction to produce methanol. Some conversion methods are more efficient, while others incur higher production costs. A moderate conversion efficiency for methanol is 0.33, but could be as high as 0.44, based on bench-scale studies of the Hynol process (a high-temperature, high-pressure method for converting biomass into methanol fuels) which has a carbon conversion efficiency of almost 87% (Norbeck and Johnson 2000).

Using the conservative conversion rate of 0.33, close to 92 M tonnes could be produced from Canadian forest-derived biomass (Table 1-1). Priced to compete with gasoline at $0.85 per litre, there is the potential to establish a market worth up to $105 billion, with the capability to displace 74% of Canadian gasoline and diesel consumption combined.

If biomass were to be used as the feedstock for methanol production, the carbon credit would be neutral. Currently natural gas is the most inexpensive feedstock for methanol production; however, advances in gasification technologies are making biomass feedstocks more desirable, especially when considering the environmental benefits and the potential for improved waste management.

Although the volume of exhaust gas produced by waste wood and clean wood is comparable, it is still better to produce methanol from clean wood, due to potential release of contaminants such as heavy metals and chlorine from waste woods.

2.1   Research and Development Required

The disadvantages of using methanol are that it is toxic, has a degree of corrosiveness and is adversely affected by temperature. Whereas methanol is the fuel of choice for racecar drivers, engines need to be flushed with gasoline during periods when it is not used, and after races to avoid corrosion and to prevent it from gumming up components. Marketers believe that fuel conditioning could solve this problem.

According to the chairman of Daimler Chrysler, methanol is only 10% more expensive than gasoline, based on European fuel prices, when considering the net fuel efficiency (Woman Motorist 2003). However, methanol may have a role in hydrogen-powered vehicles. A hydrogen fuel tank would take up the entire trunk space of a car, while methanol can be stored in regular fuel tanks, catalyzed into hydrogen within a cell and then burned. Unfortunately this system still emits 2/3 the carbon dioxide (CO_2) emissions of an internal combustion engine (Woman Motorist 2003).

3.   BIO-OIL

The petroleum refining industry has provided a historic example of how to maximize the economic value from crude oil. This same model can be applied to the "bio-refining" business, producing a renewable, CO_2-neutral "bio-oil" crude that is friendly to the environment.

The refining of biomass using fast pyrolysis technology (*i.e.*, the conversion of organic material using heat, essentially without combustion or incineration) yields a liquid bio-oil suitable for use in boilers and stationary turbines. One of the advantages of bio-oil is that it can be stored and transported, whereas gasification and combustion of biomass typically requires it to be used immediately upon production (Ensyn Group Inc. 2000).

To produce bio-oil, wood or other biomass is fed into a heated vessel, which is contacted with a stream of hot sand. At a temperature nearing 500°C, the biomass is vaporized and when cooled, condenses into a dark liquid. The yield of liquid fuel from a typical hardwood is approximately 75% by weight. Other yields are char (15%) and combustible gases (12%). In addition, approximately 30 value-added chemicals are recovered from the refining process, which can be sold to the food, polymer, petrochemical and natural chemical industries.

As the current world leader, the Canadian company Ensyn produces and sells these natural chemical and energy products from renewable wood residues, agricultural materials and other biomass feedstocks. The finished bio-products are manufactured using a unique core technology, known as Rapid Thermal Processing or "RTP™", which adds significant value to abundant, low-value biomass (Ensyn Group Inc. 2000).

The company's principal design, engineering, and research and development facilities are located in the Ottawa area, along with one commercial facility, while three other commercial facilities are in Wisconsin. Another Canadian company, DynaMotive Energy Systems Corporation, has entered the commercialization phase of its Biotherm™ pyrolysis technology with the launch of the world's first bio-oil co-generation facility in West Lorne, Ontario. At this location, enough energy will be produced to meet the needs of the sawmill, as well as electricity to power 2,500 homes (DynaMotive 2005).

The company Renewable Oil International™ has developed mobile units for bio-oil production. For example, a 125 ton-per-day plant can be designed to fit in three shipping containers (2.4m × 2.4m × 6m) and a 2ha parcel of land would be enough to run such a plant, including space for biomass loading and storage (Renewable Oil International 2005).

From the 207 MT of available biomass in Canada, it would be possible to produce 155 MT of bio-oil, with a value of almost $40 billion (Table 1.1). On an energy equivalent basis, this would be enough to replace 43% of Canada's current energy consumption (Table 2-3).

4.   HYDROGEN

Hydrogen fuel is an especially attractive form of energy because it is a non-polluting, quiet, efficient source of electricity that can be generated from any fuel source. Although hydrogen can be produced using nuclear power, today it is obtained largely from fossil fuels, with 90% of the commercially usable hydrogen currently produced through steam reforming of natural gas (Nath and Das 2003). While some observers expect natural gas to continue to be the feedstock of choice for hydrogen producers, others point to ever-increasing prices as an indicator that renewable feedstocks will become more important. Forest biomass, which contains about 6% hydrogen, is a likely source of material for hydrogen production, and particularly attractive due to its near-zero CO_2 impact (Czernik et al. 1999).

4.1   The Conversion of Biomass into Hydrogen

There are numerous ways of producing hydrogen from biomass, including: 1) thermochemical gasification coupled with water-gas shift, 2) pyrolysis followed by reforming of fractions of bio-oil, 3) direct solar gasification, 4) various novel gasification processes, 5) biomass-derived syngas conversion, 6) supercritical conversion, and 7) microbial conversion (Nath & Das 2003).

Thermochemical gasification coupled with water gas shift currently is the most widely-practiced method. The energy conversion efficiency of a system that combines gasification of woody biomass, reforming and shift conversion is approximately 79%, including the by-product steam. From each metric ton of dry biomass, about 66 kg of hydrogen and 10 million BTU are produced (the energy equivalence of about 300 litres of gasoline).

Production costs could be reduced by using these 10 million BTU of heat produced to generate electricity, or as a thermal energy source in systems such as pyrolytic conversion of biomass to charcoal, gases and methanol. Also, biomass gasification is becoming increasingly efficient and capable of making use of municipal waste wood, logging residues and urban organics, as well as plastics, styrofoam and rubber. Gasification of biomass is most economically and environmentally favourable when it uses these low cost "waste feedstocks" and production is at a large scale (Czernik et al. 1999).

Producing hydrogen in combination with the production of bio-oil through pyrolysis and catalytic steam reforming is particularly beneficial, partly due to the opportunity of extracting co-products (Nath & Das 2003). This approach can yield hydrogen in the range of US$6-8/GJ. If all of the currently available biomass was converted to bio-oil, and hydrogen extracted from the intermediate conversion, 10 MT could be produced, with a market value of $32 billion (Table 1-1). On an energy equivalent basis, this could replace 13% of current gasoline consumption (Table 2-3).

Steam electrolysis uses energy in the form of heat (at 2,500°C) to split water molecules. If steam or thermal energy is readily available, this process can be more efficient than conventional electrolysis. Other variations include: thermochemical splitting, whereby chemicals (bromine or iodine) assisted by heat are used to split water molecules; phyto-electrochemical process, which makes use of soluble metal complexes that absorb solar energy to drive the water splitting process (similar to photosynthesis in plants); and biological and photobiological processes, using algae and bacteria to produce hydrogen.

When the first energy crisis of the 1970s prompted research into new ways of generating energy, a number of photosynthetic bacteria, non-photosynthetic bacteria, cyanobacteria, and green, red and brown algae were discovered. These organisms produce the enzyme hydrogenase, which acts as a catalyst to split water molecules, a necessary step in the production of hydrogen. Such a system could supply the world's current energy needs using 0.5 million km^2, a 0.1% fraction of the earth's surface, an area about the size of France (Schor 1994).

4.2   The Automobile Industry

Hydrogen gas can be used as a fuel in combustion engines or in a fuel cell in electric vehicles. The automobile and oil companies see conversion of gasoline to hydrogen as a promising way forward to replace gasoline engines. The US government projects that a fuel cell car will perform at 60 miles per gallon of gasoline, (3.9 L/100 km), or 87 km/kg of hydrogen (1 kg is equivalent to 1 gallon of hydrogen).

Over thirty years ago, General Motors coined the term "Hydrogen Economy," but the real push to convert to hydrogen cells is only a recent phenomenon. In 2000, the US government launched a $1.5-billion initiative to convert transportation to hydrogen cell technology. At about the same time, the European Commission began to fund the European Integrated Hydrogen project, a 20-industry member effort to harmonize regulations and new codes for hydrogen-fueled vehicles and fueling stations (Morris 2003).

In January 2001, Toyota joined the GM-Exxon/Mobil alliance to develop gasoline-based fuel cell cars (Morris 2003). In June 2001, Nissan and Reneault announced they would make gasoline-based fuel cell cars their priority and initiated a five-year joint research and development (R&D) program supported by a nearly $900 million investment. In January 2003, they announced plans to begin limited marketing of fuel cell vehicles in Japan, two years ahead of their previously announced schedule (Automotive News 2002).

Today it appears that previous cost projections associated with hydrogen fuel cell vehicles have been grossly underestimated, due to the high costs associated with hydrogen storage and dispensing, as well as the slower-than-anticipated rate of introduction of hydrogen vehicles (Doty 2004). From a sustainability perspective, however, it is critical to support the development of hydrogen cells that will use biomass-derived hydrogen.

4.3   Future Directions

Unlike most fuels, hydrogen is a gas at ambient temperatures and pressures, and thus requires densification into a liquid for storage and transportation purposes. There is no transportation or distribution system yet available, aside from rail and highway transit. Thus, one of the larger issues for the hydrogen economy is condensing the fuel. Systems such as compressed hydrogen, liquid hydrogen, and chemical bonding between hydrogen and storage materials (for example metal hydrides) are currently being investigated.

The expected date for the use of hydrogen and hydrogen fuel cells on a large scale basis is impossible to predict. The technology for this type of energy is nearing the market; for instance, researchers in Minnesota recently developed a prototype reactor that produces hydrogen from corn-based ethanol which is both small (60 cm high) and efficient enough to power a car or heat a home, and inexpensive enough for consumer use (Deluga et al. 2004).

Transition to a hydrogen economy also will require an infrastructure conversion process of epic proportions. Storage technologies, electrolysis, hydrogen fueling stations, fuel cells, vehicles and more are all under development and need to mesh prior to their emergence. Government involvement will be needed to aid in establishing both the infrastructure and a price-competitive market for launching hydrogen energy and fuel cells.

5.   WOOD AS A PRIMARY ENERGY SOURCE

Close to one-third of Canadian households have wood-burning equipment. According to Statistics Canada, 3 million m^3 of roundwood was harvested in Canada in 1999 for fuelwood and firewood (Natural Resources Canada 2000), with an additional large amount being acquired rather than purchased. The primary energy value of the wood used for residential space heating is about 90-96 PJ/year (Natural Resources Canada 2002).

Overharvesting for fuelwood is not presently an issue in Canada. However, low thermal efficiency and widespread air pollution are negatives associated with burning of loose biomass in conventional devices. The burning of raw material in conventional stoves has a low conversion efficiency with up to 40% particulate emissions in the flue gases (Regional Wood Energy Development Program 2002). Adding to the wastefulness is the large percentage of unburned carbon in the form of ash. Fortunately, technological advances in catalytic combustion over the past decade have made wood stoves more efficient and particulate emissions have been reduced by about 80%.

It is important to understand and monitor wood moisture content, especially if latent heat is not being recovered. Green wood is high in moisture and its net available energy content is 3,230 BTU/lb (7,513 kJ/kg) (JUCA Super Fireplace 2003). However, when cut to length and seasoned, its moisture content will drop to 20% and have a heat value of 6,080 BTU/lb (14,142 kJ/kg) (JUCA Super Fireplace 2003). In an industrial setting where latent heat can be recovered, the high heat value of 6,930 BTU/lb (16,119 kJ/kg) is used. In some plants that are implementing turbines for electricity production, latent heat is used for hydronic space heating or combined with steam for processing.

5.1   Wood Chips

Woodchip combustion is a feasible alternative to fossil fuels but it is not widely used in Canada. Woodchips can be burned in a highly efficient combustion chamber and the energy generated used to heat water. In a domestic setting, heat is provided by hot water that circulates in pipes throughout the home. In commercial operations, steam is produced to power a turbine, which generates electricity.

While the wood processing industries use much of the waste wood that they generate, surplus amounts can be used for district energy systems. In communities where timber and pulpwood are processed, it is possible to use the waste industrial heat in district energy systems to heat large buildings and even residential homes that are nearby. Two communities in British Columbia, Revelstoke and Masset, are studying this option (International Energy Association 2005), and three others, namely Charlottetown (Prince Edward Island), Oujé-Bougoumou (Quebec) and Grassy Narrows (Ontario), currently use surplus wood waste from nearby sawmills for this purpose.

Charlottetown has the largest biomass-based district energy system (1.2 MW) in Canada. It is fuelled by a combination of municipal solid waste (45%) and sawmill residue (45%), with only 10% generated by oil (International Energy Association 2005). The combined municipal waste and sawmill residue displaces about 17 million litres of heating oil per year. Depending on moisture content, a tonne of woodchips can replace 169-300 L of oil (Table 2-4).

Woodchip or biomass combustion is especially helpful in "energy islands", a term used to describe places where energy is difficult or costly to acquire. There are some 150 remote communities in Canada that don't have access to natural gas pipelines or electricity grids, yet have significant forest resources. Pilot studies have indicated that bioenergy-fired mini-district heating systems would be an ideal alternative to expensive, imported oil for these communities (McCallum 1999) and also would provide opportunities for sustainable forest management practices and long-term employment.

Table 2-4. The Effect of Moisture Content on Energy Value of Woodchips.

Moisture Content	Displaced Litres of Oil/Tonne of Woodchips	Net Amount of Energy (GJ)[1]
30%	300	11,582
35%	273	10,540
40%	247	9,536
45%	211	8,146
50%	195	7,529
55%	169	6,525

Source: McCallum 1999 (assumes 80% system efficiency for oil burners and 70% efficiency for woodchip combustors).
[1]36,592 BTU/L for heating oil from Appendix A (Chapter 1).

The estimated cost for delivered wood chips is $50-70/ton, depending on factors such as the size of the operation, equipment used and biomass source (McCallum 1999). These estimates are based on stand-alone chipping operations; however, costs could be reduced if chipping and logging operations were fully integrated. Combining pre-commercial thinnings and logging slash (management techniques to reduce the risk of forest fires) also could help to decrease overall costs.

For Canada as a whole, logging residues are immense. If all 75.8 M tonnes were processed as wood chips, it could establish an economy worth up to $9 billion (Table 1-1). If biomass from all sources was used (182.4 MT), it would have an energy value equivalent to 26% of Canada's annual energy production (assuming 100% energy gain and fuelwood availability), or 51% of energy consumption (Table 2-3).

One of the problems to be overcome in the use of woodchips is the high cost associated with transportation and switching to new heating systems. In addition, there needs to be regulation and quality control of chips (since the thermal properties of most biomass can vary, a monetary value would need to correspond with a heat rating, like BTUs or degree hours). Co-ordination between chip suppliers, handlers and buyers would be required. In addition, public insecurities over converting to new technologies and security of supply would have to be addressed. Many of these problems could be solved if there was sufficient motivation to scale up operations to develop a feasible system and to gain public confidence.

5.2   Charcoal

Charcoal can be produced from wood and other types of high fibre biomass. Traditionally, charcoal was made by burning large piles of wood while watering the surface to create an anaerobic environment. Today, charcoal is produced in kilns through pyrolysis, where chemical bonds are broken at high temperatures in the absence of air and re-form at higher density. During this process, water and volatile hydrocarbons are driven from the wood and the resulting product is a highly efficient, calorie-rich material that burns hot enough to smelt steel. In terms of transportation, it is more cost-efficient to distribute charcoal than wood. If done properly, with sufficient demand, logging residues and municipal waste wood could be converted to charcoal with the potential for gas extraction (as a measure to decrease emissions and to recover potential energy) and further processing into briquettes/pellets.

5.3   Briquettes, Pellets and Densified Fuels

Pellets and briquettes are generally made from compressed sawdust, ground wood chips, shavings and other waste materials from the manufacturing of furniture, lumber and other wood products. Briquettes, which are larger than pellets, may also contain charcoal. The resins and binders that occur naturally in sawdust (particularly softwood) help hold wood pellets together. Bark and logging waste also make durable pellets; however, the ash content is slightly higher and they have a lower heating value (Lehtikangas 2001).

Other fuel products include fire logs, fuel-penetrated fire starters and barbecue briquettes. Their contribution to the bioenergy sector, although marginal, does assist in supplementing fossil fuel-derived energy and decreasing CO_2 emissions.

Some of the advantages of these densified fuels are that they can be generated from a range of materials with varying dimensions, they have a low bulk density, smaller space requirement (lowering transportation costs), and their low moisture content means they provide a much higher BTU output per pound than firewood or woodchips. Commercially produced pellets have a moisture content of 10-12% (Lehtikangas 2001), and a density of 1.2 g/cm^3 (compared to loose biomass with a bulk density of 0.1-0.2 g/cm^3). For an equivalent energy output, the cost is $13.71 for softwood pellets, $30.80 for propane and electricity, $16.29 for oil, and $20.00 for natural gas (Dansons Group Inc. 2003).

For home use, pellets and briquettes require less storage space than cordwood and burn much longer, thus requiring less effort for managing the fire. They are clean (absent of dirt, leaves, insects and water) and can be poured into a fire without stoking. Specialized stoves are also available that have oversized hoppers, or engineered self-feeding augers that can continue to run for days without refueling.

Using all of the currently available biomass, 273.85 MT of briquettes could be produced with a potential market value of over $80 million (Table 1-1). The commercial success of briquetting is limited by the understanding of the biomass itself, technological advancements in machinery, and the availability of stoves that can draw out the most heat and produce the fewest gases and ash. Softwoods have a low density and so transportation is more costly than for hardwoods; however, if transportation costs are reduced by production near the source of these biomass materials, pelletization can be an effective way of preparing wood-derived fuel.

6.   BIODIESEL AND OTHER EMERGING TECHNOLOGIES

6.1   Biodiesel

Studies are underway to determine whether forest biomass-derived diesel production could be both technically and economically feasible, either as a substitute for transport fuel or as a blend through the BIOCAP Canada (Wood and Layzell 2003). Government incentives likely will be required, such as imposing the use of a certain percentage of biodiesel by a set date as is being explored with ethanol, along with tax incentives, capital investment for a new fuel supply infrastructure, and funding the necessary research and development.

Canada has been experimenting with biodiesel made from agricultural products in Saskatchewan since 2001. The 500,000 L produced at the plant are being tested as fuel conditioners and marketed across Canada (Agriculture and Agri-Food Canada 2002). While biodiesel derived from agricultural crops has the advantages of high energy content, low emissions and the ability to make marginal land productive, agricultural crops alone cannot displace petroleum. If, for example, a 2% consumption of biodiesel became mandatory, this would initiate a market for 468 million L by 2010 (total consumption of diesel in the Canadian market in 2000 was 23,380 million L). Even if all of the 35 M ha of cropland in Canada were dedicated to soy cultivation, there would still be a demand for about 2,217 million L. If biomass-to-biodiesel production were feasible, farmland could be put to use for other purposes. As with ethanol marketing, it is necessary for the government to initiate incentives, rally consumer support, and provide funds to support research and development.

The feasibility of using biodiesel in Montreal's transit buses for a one-year period was tested starting in March 2002 and was known as the BIOBUS project. Results showed that using biodiesel had very positive effects, such as superior lubricity, improved ignition performance and lower emissions, even under cold winter conditions where temperatures dropped to −30°C (BIOBUS Project 2003).

6.1.1 Current Costs of Biodiesel

In 2001, the cost of B100 (100% biodiesel) was about US$3.00 per gallon (US$.79/L) (Association of Equipment Manufacturers 2001). More commonly used is the B20 biodiesel blend, which costs an extra US $.20-$.40 per gallon (US$.05-$.11/L), compared to petrodiesel. For example, on the West Coast in July 2002, the price for the B20 biodiesel blend was US$1.79 (US $.47/L) compared to US$1.38 for a gallon of petrodiesel (US $.36/L); however, according to the US Department of Energy, people who can afford it are buying it because they feel it is the right thing to do.

6.1.2   Technical Considerations

Since biodiesel is produced instead of being refined, its properties can be altered for optimal performance in a variety of engines. Diesel molecules are fairly simple, and to meet specified criteria, require the following: the carbon chain should be 45-60 carbon long, they should have a high lubricity, contain low sulphur, and have an energy value in the range of 117,000 - 130,000 BTU/gal (32-36 GJ/L) (McCormick 2003).

Methyl ester, ethyl ester, and hydrogenated soy ethyl ester (HySEE) are a few types of biodiesel that are currently being tested for various properties, such as flash point, cloud point, combustion and viscosity. Biodiesel can be used alone or mixed in any ratio with petroleum diesel. As of 2001, at least seven companies were marketing diesel with increased lubrication by adding biodiesel in small quantities (Association of Equipment Manufacturers 2001).

Despite the power loss from which some engines suffer (primarily older models), most engines respond positively to biodiesel. Forthcoming diesel engines can be configured specifically for biodiesel, to gain the most power possible. Engines fueled with biodiesel emit less visual smoke, have equal or better engine wear, and are brighter and cleaner internally (Peterson et al. 1999).

Some concerns identified illustrate areas where further research will be needed. These include the flow problems during cold-weather operation; B20 blends exhibit gelling and cold flow problems 3-5°F (2-3°C) before conventional diesel. Also, stability of biodiesel appears to be decreased by the high oxygenate content that can cause varnishing, gumming and plugging. Biodiesel can soften and degrade elastomers and natural rubber, and can damage fuel lines, hoses and pump seals. Further work will be required to research compatible corrosion-inhibiting additives, elimination of deposits, and the development of anti-oxidants to increase storability and standardize shelf life. Duration testing in larger engines will be required as well as biodegradability determination.

6.1.3   Environmental Considerations

Biodiesel reduces CO_2 greenhouse gases by 78% over its entire life cycle (National Renewable Energy Laboratory 1998) and has a positive energy balance of 3.2:1, while regular diesel fuel has a balance of 0.83:1 (Association of Equipment Manufacturers 2001); thus, it is increasing in popularity as an environmentally sound renewable resource. In addition, the production of biodiesel requires far less technology than both ethanol and petrodiesel.

The Environmental Protection Agency (EPA) in the US intends to require refiners to reduce the sulphur content of diesel, which causes extra wear on mechanical parts and pollutes the air, to 15 ppm on-road and 500 ppm off-road by June 1, 2006. The projected increase in cost for 15 ppm sulphur diesel (estimated to be $.13 per gallon or $.34/L (Association of Equipment Manufacturers 2001)) will make biodiesel more competitive since it contains a fraction of the sulphur.

6.2   Fischer-Tropsch Diesel

Fischer-Tropsch diesel (FTD) is a high quality diesel fuel that has been used since World War II. It is a synthetic fuel that is produced by condensing natural gas (methane) into a liquid by the costly "gas to liquid" (GTL) refining process. It can be generated via biomass gasification in the presence of steam and oxygen to produce CO and H_2, which are then converted to paraffin and then to diesel by the process of FTD-synthesis. The GTL conversion process can use any hydrocarbon-rich material as feedstock (Association of Equipment Manufacturers 2001). It can be economically feasible to make FTD if the gases are by-products and essentially free, as in biomass processing. The near zero sulphur content, elimination of aromatics and a high cetane value (measure of the ignition quality of fuel) collectively, make GTL conversion fuels superior in many ways. In an attempt to lower gas emissions, California is investigating the full potential and economic feasibility of GTL fuels.

The German company CHOREN, working with Daimler Chrysler and Volkswagen, has made progress in producing FTD from wood and biomass waste products, and elsewhere in Europe, a facility has successfully produced FTD from willow (Greene 2004).

6.3   E-Diesel

Laboratory and field tests are currently being conducted on Ethanol-diesel (E-diesel), a blend of ethanol and diesel. E-diesel blends are noted for measurable improvements in exhaust emissions. Ethanol and diesel fuels do not mix readily; thus a new fuel delivery system is required. Still in its infancy, there are performance issues yet to be addressed. In a 10% ethanol blend, power output decreases and fuel consumption increases by 4-5%. This is due to the relatively low energy content of ethanol. Another significant drawback is that ethanol decreases the lubricity of diesel. Despite the numerous obstacles, research and development continues in this field due to fuel tax incentives and the pressing need for renewable biofuels. Economic estimates suggest that a 15% E-Diesel blend could cost about 5-7% more than conventional diesel (Association of Equipment Manufacturers 2001).

6.4   Ethyl Levulinate

Ethyl levulinate, produced from Levulinic acid through esterification, has properties similar to some fuels. Preliminary studies have shown that E-levulinate has good lubricity, flashpoint stability in blends up to 10%, and similar cetane numbers. As an alternative to diesel fuel, it is less expensive and cleaner burning (Bio Development International 2003).

6.5   BtL Fuel

A new biobased fuel, biomass to liquid (BtL fuel) is receiving increased attention in Germany (Presusser 2004). It can be synthesized from waste biomass, such as wood chips, and uses gasification and Fischer-Tropsch technologies in its production. Shell has already introduced a "V-Power diesel" formulation in the German market to prepare the consumer. It is expected that some European countries, such as the Netherlands, will need foreign sources of biomass to supply their plants with an estimated 15 million tonnes of biomass per year. While transport of wood chips in such large quantities would not be feasible, preprocessing biomass into bio-oil using the fast pyrolysis technologies that Ensyn or DynaMotive have developed may provide Canada with a role in this type of endeavour.

6.6   Biocrude

Many plant species native to Canada contain extractives that have similar qualities to crude oil and petroleum fuel. Plant hydrocarbons such as α-pinene, β-pinene and heptane, in lesser amounts, can be cracked like oil to produce biodiesel. Knowledge of these hydrocarbons, also known as biocrude in the raw form, broadens the realm of possibilities for biofuels. Biocrude production provides a raw material for bioplastics, platform chemicals, gasoline, diesel, fuel oil, latex and rubber. In addition, biocrude has lower sulphur, nitrous oxides and heavy metals than fossil crudes, making it safer to handle and less harmful to the environment.

While oil seed plants in agriculture (such as peanuts, safflower, sunflower, canola, soybean, corn flaxseed and cottonseed, which produce in the range of 140-800 kg/ha of oil) convert carbohydrates into fatty acids and glycerides, evergreens and highly resinous plants convert them into hydrocarbon chains such as terpenes and latexes. Turpentine and other cyclic terpene alcohols were extracted from pondersoa pine (*Pinus ponderosa*) and used as a source of combustible fuels in the early 1800s.

Terpenes can be treated like oils and with simple chemical modifications, they can be converted into biodiesel (Weisz and Marshall 1979). Most plants and trees presently under study originate from tropical locations. If biofuels are to be successful, additional research is needed to explore and catalogue the abundant hydrocarbon chains (such as terpenes, latex and oils) that exist within the diverse taxa of North America.

Coniferous bark and needles are oil-rich, and by harvesting crown material from softwood, oils and extracts could likely be used for biodiesel. If all the terpenes were extracted from disposed Christmas trees, as well as from conifer boughs from logging, some amounts of biodiesel could be produced. Moreover, if the waste biomass was pelletized and used to generate electricity, the energy produced could be used for the plant or in a district heating system.

To exemplify this, moleplant (*Euphorbia lathyris*) (a plant under investigation for biofuel production) is composed of 8% terpenes for biofuel production, 20% as fermentable sugars and 30% lignocellulose that can be used as a combustible to eliminate fuel consumption in the conversion process. Based on Calvin (1985), *Euphorbia lathyris* yields a dry biomass of 21.3 T/ha and is capable of producing 0.58 tonnes of biocrude and 0.68 tonnes of ethanol per ha. If the entire 1500 km^2 (Hydro One has 29,000 km of lines + 1000 km for Great Lakes Power and others = 30,000kms at 0.05 km wide) under the Ontario transmission lines were used to cultivate *E. lathyris*, 214,803 and 252,381 tonnes of biocrude and ethanol, respectively, would be produced with a combined market value of $190 million. Realistically, land easements, accessibility and terrain limitations will significantly decrease the land available for production; however, this example using a small portion of wasteland illustrates the scale on which biomass could supplement petroleum.

6.7   Oil Production from Algae

Petroleum is believed to have originated from kerogen, which can be converted into oily substances under high pressure and temperature, paralleled with oil production through pyrolysis. Kerogen is formed when algae, biodegraded organic compounds, plankton, bacteria, plant material and other substances undergo biochemical and chemical reactions known as diagenesis and catagenesis. Recent findings indicate that fungi and sludge can be converted to oil under elevated temperatures and enriched CO_2 conditions. The same may be possible with algae.

The remarkable algae *Botryococcus braunii* is unique in that it produces large volumes of oil, up to 86% of its dry weight (Calvin 1985; Miyamoto 1997). The oil derived from *Botryococcus braunii* is so similar to crude oil that researchers suspect it has its origins in the world's great oil fields. This organism grows photosynthetically on treated sewage, and while there is no indication that biomass has been used as a culture medium, the nutrient composition of *in vitro* culture suggests that biomass could serve as a raw input to augment the growth potential at low cost. Chemical analysis of biosolids at St. Mary's Paper (a thermochemical paper mill in Sault Ste. Marie, Ontario, Canada) indicates that macro- and micronutrients used to culture *B. braunii* in lab media are present in the biosolids. Furthermore, although the algae are photosynthetic, cellulosic material could be converted to reducible sugars as in ethanol production to accelerate growth rate and subsequent biocrude production.

These oil-rich microorganisms are photosynthetic and therefore combustion emissions would be partially offset by net CO_2 absorption. Although *Botryococcus braunii* has been more thoroughly studied, other oil-producing algae possess outstanding characteristics. *Nannochloris* species, which also have high lipid content and produce voluminous amounts of triglycerides, can serve directly as diesel fuel.

Efforts to cultivate algae in bioreactors and on various growth media have proven successful. Harvesting techniques are simple. The algae are grown in large tanks, float to the surface after reaching oil-induced buoyancy, and are skimmed and lysed to yield a biocrude. The oil can be upgraded to petroleum products via standard oil refinery processes. The benefit from biocrude production in bioreactors is that the various factors that influence growth can be easily manipulated. In fact, extrinsic manipulation can alter the content and proportions of glycerol, poly-unsaturated and saturated fatty acids. This system could be more promising than that of ethanol, which involves many steps, is costly and has relatively low yields (~20% from wood biomass). However, there is a cost associated with the nutritional supplementing of the growth media. Thus the use of slurries deriving from wood wastes, papermill sludge and other biomass could be the key to establishing algal biocrude production at a competitive price.

Future work is required to identify the variety or strain of algae that yields the highest oil quantities and can use pulp, paper or wood mulch as a carbon source, as well as to investigate and catalogue organisms that convert wood and organics to oil, to further study organisms that convert sugar to oil, and to develop a refinery process similar to ethanol production (cellulose to sugar to oil).

7. NOVEL WAYS OF PRODUCING HEAT

In 2000, residences consumed 17.5% of total energy demands, and it is estimated that more than 80% of residential energy is used for space (60%) and water heating (20%). Assuming that space heating constitutes 50% of the $27.6 billion spent on utilities Canada-wide in 2000 (Statistics Canada 2002), approximately $13.8 billion was spent on space heating. This market is large enough that minimal improvements would have a large impact on energy demands and could make a serious contribution to reducing GHG emissions.

Presently, there is technology available in Canada to utilize biomass-derived fuels in high efficiency devices for space heating. Wood chips, briquettes and pelletized material are an inexpensive feedstock to fuel boilers. Hydronic heating, for example, which is 5% to 15% more efficient than standard forced air systems, is gaining popularity and typically draws heat from boilers, which can be fuelled with biomass.

Regardless of advances in space heating and development of high efficiency furnaces and boilers that operate at 90% and higher, CO_2 will forever be a by-product of fossil fuel combustion. Heat recovery and transfer will have to be implemented to maximize energy gains from biomass.

7.1 Extracting Heat from Microbial Decomposition

It would be worth investigating how much heat could be extracted from biomass undergoing microbial decomposition. Knowing that material conducive to fermentation can generate temperatures high enough to spontaneously combust (Mehaffey et al. 2000; Hutton 2002), it is theoretically possible to use this heat source in a hydronic system. If piles of waste biomass were stationed near institutions, factories or set up as district heating systems, heat could be harnessed and transported through a network of water-filled conduits and used as a source of radiant heat. For household use, biomass could be set in an insulated vessel with coils or baffles running through it. Water could then be heated and pumped throughout the home. Energy could also be saved by boosting cold water by 2-3°C prior to entering the hot water tank.

Ideally, such a system would work concurrently with a gas collector. Harvested methane could be used to run devices such as furnaces for back up heat, hot water tanks and other household devices. Set up as a district heating system, thermal energy from biomass could maintain pipeline temperature at regular intervals and during peak demands, gas boilers run on methane by-products or cogeneration plants fueled with waste wood could kick in.

7.2 Generating Methane from Biomass

Natural gases extracted from the ground increase net atmospheric CO_2, while methane harvested from naturally fermented sources such as municipal solid wastes infer a source:sink balance. Natural gas is heavily used by Canadian industrial and commercial sectors and is the primary mode of space heating. In 2001, Canada produced 195 billion m^3 of natural gas, and 67 billion m^3 of this was consumed by Canadians. The total amount of Canadian natural gas remaining in reserves as of 2000 is 1,683 billion m^3 (Natural Resources Canada 2000a).

Anaerobic digestion of wastes, a simple process that occurs naturally and takes place in the absence of air, results in the conversion of organic matter, by bacterial action, into a mixture of methane and carbon dioxide. In this process, over 93% of the effluent is converted to gas, leaving 3% as sludge, which is more efficient than the comparable biomass conversion to alcohol (Schor 1994). Currently, feedstocks include municipal solid wastes, yard trimmings, garden wastes, agricultural residues and animal wastes. BioTechnica, a US firm, has been examining methods by which landfill waste-disposal sites can be converted into "bioreactors" for methane production. Many landfills take in 5,000 tons of refuse every day. It was estimated that this type of process could satisfy 1% of the US national energy requirement (Schor 1994).

With infrastructure and trans-Canadian transmission system in place, it should be straightforward to convert to biomass- and waste-derived methane. On-site production would make it a versatile fuel for use in remote locations, distant from any pipelines, and very cost efficient, especially when generated from wastes.

7.3 Gasification Technologies

While biological systems can produce methane, thermochemical treatments can yield the same, and in less time. Organic materials are transformed into gases rich in methane, small quantities of liquid including methanol, and a solid residue containing carbon and ash. The off-gases may also be treated in a secondary thermal oxidation unit. There are several types of pyrolysis units including rotary kilns, rotary hearth furnaces or fluidized bed reactors, which are called parallel gasification technologies.

The integrated biomass gasification combined cycle converts biomass to gaseous fuel, comparable to natural gas, for combustion in gas turbines with heat recovery to run a steam turbine and/or for generating electrical energy. Heat recovery and recycling of by-products result in high thermal efficiencies, making this system superior to conventional technologies.

Presently, utilizing waste material can generate power at a competitive price; however, implementation of the technology is limited by biomass supply. Government involvement and support of bioenergy and dependant biomass procurement would facilitate the solution to this problem. What makes integrated biomass gasification and similar technologies particularly attractive is their ability to process a variety of waste materials and renewable resources, such as energy crops and forestry residues, with high efficiency and low emission levels (Krigmont 1999; Kettle 2001).

Despite lacking documentation, it is becoming apparent that gasification for syn-gas and methanol production could supersede biologically-attained fuels. Biological fermentation to yield compounds such as methane, alcohols, and platform chemicals such as levulinic acid and lactic acid may not be possible using contaminated feedstocks like municipal waste wood. Materials that are rich in carbon, such as peat moss, may not have high cellulose content, the precursor to biological transformation; however, via gasification, most carbon-backbone molecules can be converted to biofuels (Krigmont 1999). For instance, the biomass-to-ethanol conversion rate of 1:0.20 (Eklund and Pettersson 2000) is significantly lower than the conversion rate for methanol via pyrolysis, 1:0.35 (Clean Fuels Development Coalition 2003). In addition, contaminated waste wood arising from industry, demolition and deconstruction may not be suitable for biological action due to toxic inhibition or interference.

Gasification is not limited to feed types and can convert a broad array of wastes, such as styrofoam, plastics and rubbers, into methane or methanol (Reichenbach de Sousa 2001). Whereas moisture content is a significant factor in the combustion of biomass fuels, conversion by gasification does not require biomass to be dry. This is important with regard to utilization of peat moss, pulping sludge and other high-moisture biomass that would otherwise require expensive dewatering prior to processing. Peat moss has the greatest potential for pyrolytic conversion to methanol and methane. Carbon-rich peat has low cellulose content and a relatively low energy density, so biological conversion to fuels or heat extraction is less efficient.

8. CONCLUSION

Canada has the potential to be a world leader in the use of renewable energy. Our vast forest resources provide the ideal opportunity to develop alternative forms of energy based on biomass. However, the distances between the source of raw materials and markets present an economic challenge. Ultimately, Canada is in search of a number of fuels that will be useable in a broad variety of applications, be cost effective, be storable without degradation and have sufficiently high energy density to be transportable at low cost. Research into the technologies associated with biofuel production, as well as government involvement in transition to their use, is imperative. Commitment to overcome the technical challenges in biofuel technology is required to reduce our dependency on oil. Cooperation from automakers will also be required.

Chapter 3

BIOCHEMICALS

Highlights and Fast Facts

⛏	Biomass has the potential to replace many valuable industrial chemicals currently made from non-renewable feedstocks.
⛏	Levulinic acid, a valuable platform chemical with a potential market of $800 million, can be made from waste paper.
⛏	Two promising new products, MTHF (a gasoline extender) and DALA (a biopesticide), have the potential to increase the demand for levulinic acid to an estimated 91,000-181,000 tonnes/yr.
⛏	Research is underway to use lignocellulosic feedstocks for the production of polylactide (PLA), a prime source of bioplastics.
⛏	Lignins, a byproduct of the paper industry, have a potential market value of $400,000 million.

One of the most compelling reasons to make the transition to a biobased economy is the enormous potential of renewable organic feedstocks to provide much more than just energy. Unlike other renewable sources such as solar, wind or tidal energy which produce electricity and heat, biomass can be converted to liquid fuels as well as supplying the basis for many of the essential chemicals and materials that are the underpinnings of our industrialized society.

Biochemicals will no doubt play a central role in the development of a biobased economy. Derived from biomass, or created through a less energy-intensive or non-polluting process referred to as "green chemistry", biochemicals have particularly significant potential as replacements for petrochemicals. Canada annually produces 27.2 million tonnes of fossil

61

fuel-derived chemicals (Industry Canada 2004), worth over $10 billion (Crawford 2001).

Many of these are organic chemicals could be made from cellulose and lignin instead of fossil fuels. In addition, some are platform chemicals, meaning that they can be transformed by methods such as hydrogenation, esterification, or fermentation, into other valuable chemicals with a wide variety of uses. Alone they may not be in a form that is useable, but further processing by refineries can produce bioproducts that have significant economic potential.

Biochemicals have a potential market worth of $1.7 billion in Canada (Crawford 2001). The estimated world market value of platform chemicals derived from lignocellulosic feedstocks is outlined in Table 3-1.

Table 3-1. Potential Market Value of Platform Chemicals.

Chemical	World Demand in Tonnes ('000)	Potential Market Value ($million)
Levulinic acid & its derivatives	91 – 181	800
Lactic acid	1,400	>10,000
Succinic acid	>32	>300
Lignosulfonates	1,000	400

Sources: Agriculture and Agri-Food Canada (2002); Oldroyd, D. North American Marketing Manger, Tembec Inc. per. comm. 2004.

The use of residual biomass for the production of platform chemicals is not only a less expensive alternative in the replacement of petrochemical products, it also provides a sustainable way of disposing of waste and adds income to the industrial and rural sectors (where communities are close to the source of feedstock). When some of the technical difficulties in the efficient production of cellulose from glucose are overcome, biomass will be a more cost-effective feedstock than the starches currently being used (Archambault 2004). Much of the glucose is lost currently during the transformation process and in some cases products are highly diluted in large amounts of aqueous streams (Archambault 2004). Research into clean fractionation, whereby lignin and hemicellulose can be dissolved with organic solvents, will help make cellulose a more viable feedstock option.

The market potential of platform biochemicals will undoubtedly grow as oil prices and the cost of waste-disposal rise. The forestry sector has the potential to reap huge economic benefits as this new era of commodity chemical production begins.

Many platform chemicals have a dual role. Ethanol, for example, is useful as a fuel in its raw form and also acts as a precursor to many industrial chemicals. Examples of cellulosic-based platform chemicals and their synthesis pathways are outlined in Table 3-2. While the use of forest-based feedstocks for the production of succinic acid and lactic acid is not yet commercially feasible, the potential exists with further research.

Table 3-2. Examples of Platform Chemicals, Their Derivatives and Potential Uses.

Raw Material	Process	Platform Chemicals	Biochemical	Potential Uses
Wood	Hydrolysis	Ethanol	Ethylene	Precursor to many industrial chemicals
Waste cellulose Waste paper	Hydrogenation	Levulinic acid	MTHF	Fuel additive
			DALA	Pesticide
			Tetrahydrofuran	Solvent
			Alpha-angelicalactone	Fuel extender
			1,4 butanediol	Polymer intermediate
			succinic acid	Specialty chemical
			Alpha-methyltetra-hydrofuran	Solvent, chemical intermediate, fuel additive
			Ethyl levulinate	Diesel fuel
			Sodium levulinate	Antifreeze
Levulinic acid	Hydrogenation	Valeric g-lactone	1,4 pentanediol	Solvent
	Dehydration	1,4 pentanediol	1,3-pentadine (piperylene)	Synthetic rubber
	Condensation	Levulinic acid + phenol	Diphenolic acid	Resins
Cellulose (glucose)	Fermentation	Lactic acid	PLA	Bioplastic
			Lactate esters	Specialty chemicals
			Ethyl lactate	Solvent, Degradable plastic polymers
Cellulose from wood waste (glucose)	Fermentation	Succinic acid	PVP	Pharmaceutical toiletries, paper, beverages, detergents
			Itaconic acid	Polymeric fibre blends

1.   LEVULINIC ACID

Levulinic acid conventionally is derived from refined petroleum but technological advances now permit the production of levulinic acid from biomass. Suitable feedstocks from the forest industry include pulp and paper mill residues, sawmill and logging residues, and solid municipal waste. It takes approximately 2 kg of cellulose to produce 1 kg of levulinic acid (US Department of Energy 1998). Obtaining enough raw materials to support the production of levulinic acid from biomass is not a foreseeable concern; Canada currently produces over 75 million tonnes of logging residues annually which, in turn, could produce 19 million tonnes of levulinic acid. As an example, if all sources of available biomass were used, assuming wood-derived biomass contains 50% cellulose, with a conversion ratio of 0.25, 274 million tonnes (a value of $77.6 million) could be realized (Table 1-1).

Through a high-temperature, dilute-acid hydrolysis process, starches, cellulose and hemicellulose can be converted from simple biomass into monomeric sugar streams and eventually into levulinic acid. This valuable platform chemical is highly reactive and very versatile. It can produce 100-200 possible derivatives that are of use to industry. The main uses of levulinic acid include:

1.   Methyl tetrahydrofuran (MTHF), a gasoline extender.
2.   Diphenolic acid, used for epoxy resins.
3.   Tetrahydrofuran, a solvent.
4.   1,4 butanediol (a polymer intermediate).
5.   Succinic acid (a specialty chemical).
6.   Delta amino levulinic acid (DALA), the active chemical in a new group of herbicides and pesticides (US Department of Energy 1998).

In addition, both ethyl levulinate, an alternative to diesel fuel, and alpha-angelicalactone, suggested for use as a fuel extender, can be synthesized from levulinic acid. Alpha-methyltetrahydrofuran has also been prepared on commercial scale and used as a solvent and a chemical intermediate. It is particularly attractive as a fuel additive because it resembles the antiknock agent methyl t-butylether (Ghorpade and Hanna 1997).

Hydrogenation of levulinic acid can produce other useful chemicals. Valeric-g-lactone, which is an effective solvent with extensive uses, can be obtained in very high yields. This compound may be hydrogenated to 1,4-pentanediol, which upon dehydration yields 1,3-pentadine (piperylene). Piperylene is known to polymerize to a rubbery mass and is therefore a source of synthetic rubber (Ghorpade and Hanna 1997).

Diphenolic acid, the condensation product of levulinic acid and phenol, is useful in the preparation of various resins. Salts of levulinic acid are water-soluble and have several types of uses. There is a possibility of using sodium levulinate as an antifreeze ingredient as a replacement for ethylene glycol, which is toxic and petroleum-based (University of Nebraska 2005).

1.1 Market Potential

There is currently a worldwide market for levulinic acid of about 500 tonnes per year, sold to the specialty chemical market for US$10,000/ton (US Environmental Protection Agency 1999). Two new promising products, MTHF and DALA, have the potential to increase the demand for levulinic acid to an estimated 91,000-181,000 tonnes/yr (US Department of Energy 1999).

A US company, BioMetics (formerly Biofine), has developed a state-of-the-art process, the "Biofine Technology", which converts waste paper that cannot be recycled into levulinic acid. BioMetics expects to produce levulinic acid for as little as US$.71/kg (US Environmental Protection Agency 1999) compared to the US$2.20/kg it costs to produce it from fossil fuels (Fahy 1996). The BioMetics processing plant can convert 2,000 tonnes of paper per day into 75 million L of levulinic acid per year (Fahy 1996). Similarly, manufacturers expect to be able to produce the fuel extender MTHF from levulinic acid for only US$.13/L.

2. LACTIC ACID

Lactic acid can be derived from cellulose (glucose) and can produce polylactide (PLA), a very versatile, biodegradable polymer. There is much interest in the use of PLA in the plastics industry. PLA is currently being made from corn, but research indicates that the process could incorporate wood biomass. If lactic acid could be made from low cost biomass, it would provide access to opportunities for direct substitution for petrochemicals (Lipinsky 1981).

The production of PLA requires hydrolysis of cellulose to yield simple sugars which are then turned into lactic acid through the process of bacterial fermentation. The monomers of lactic acid, once achieved, are treated with chemicals to get them to link up into long chains or polymers, which bond together to form a material called polylactide.

One of the most desirable characteristics of PLA is that it breaks down easily and will lose all mechanical properties when composted for five weeks (University of Nebraska 2005a). Polylactides have a wide range of everyday uses, from garbage bags to fast food containers. While they are currently two or three times more expensive than conventional plastics, they hold enormous potential due to this ability to biodegrade and the fact that they are made from renewable feedstocks.

In addition to biodegradable plastics, PLA can be manufactured into other useful products. Researchers are currently developing technologies for the use of PLAs for the purpose of time-releasing herbicides and insecticides. The polymers are produced in such a way as to contain the active agents of biopesticides. Over a period of weeks in the field, the polymer breaks down, thereby releasing the herbicide or insecticide. Horticulturists are also interested in PLA because of the potential to make biodegradable planting containers.

The company Cargill Dow LLC produces PLA entirely from renewable resources, namely corn, but is working to develop new conversion technologies to facilitate the use of lignocellulosic feedstocks from agricultural wastes (Vink et al. 2003). Their plant in Blair, Nebraska, the world's only commercial production facility for PLA, expects to produce 140,000 tonnes of PLA annually (Cargill Dow 2004). A wide variety of products can be produced, allowing Cargill Dow to modify PLA for a range of applications such as carpets, shirts, bottles, cups, wraps that can be used in packaging, fibres and other emerging applications.

2.1   Market Potential

The potential world demand for PLA has been estimated by Cargill Dow LLC to be worth over US$10 billion (Agriculture and Agri-Food Canada 2002). Other market areas include fertilizers and pesticides, marine plastic, degradable coatings for paperboard, compost waste bags and agricultural mulch film.

PLAs are currently selling for US$1.50-$3.00/lb and it is expected that the price will fall to US$1.00/lb once production capacity increases (Biby 2005). Although the raw materials are relatively inexpensive, processing costs remain higher (for now) than with conventional technology.

Increasing international demand for bioplastics opens the door to the creation of new commodities and creates business opportunities from waste products generated by forest operations and processing. Consumers need to know that the cost of currently cheaper, traditional plastics does not reflect the impact on waste management, and municipalities are becoming aware of the significant savings that accepting only biobased products could bring. Public opinion and government policy are strong forces behind the worldwide acceptance of bioplastics by consumers and businesses alike.

3. SUCCINIC ACID

Research is currently underway to determine whether succinic acid could be produced economically from forest biomass. A novel bacteria (AFP111, a patented strain of *E. coli* currently used in the production of succinic acid from corn-derived glucose) is being genetically engineered in the lab to enhance the fermentation of sugars derived from wood wastes (US Department of Energy 1999a). If successful, this experimental process would allow the production of succinic acid at a lower cost, so that it could be used directly or as a precursor for many industrial chemicals. Potential uses include the manufacture of plastics, clothing, fibres, paint, inks, food additives and automobile bumpers.

4. LIGNINS

Lignin, the natural glue that holds wood fibres together, is one of the most abundant renewable raw materials and to date has been largely under-exploited. Lignins are a bi-product of the paper industry. Lignosulfonates, produced from the acid-bisulfite pulping of wood, have literally hundreds of applications, due to their dispersing, binding, complexing and emulsifying properties. A few of the current uses of lignin is summarized in Table 3-3.

Table 3-3. Uses of Lignin.

Binding Agent	Dispersant	Emulsifier	Sequestrant
Coal briquettes Ceramics Carbon black Dust suppressants Fertilizers & herbicides Plywood & particle board Animal feed pellets Fiberglass insulation Linoleum paste Soil stabilizers	Cement mixes Pigments & dyes Clay & ceramics Oil drilling muds Pesticides & insecticides Leather tanning Concrete admixtures Gypsum board	Asphalt emulsions Pigments and dyes Wax emulsions Pesticides	Micronutrient systems Cleaning compounds Water treatments for boilers & cooling systems

Source: Lignin Institute (1992).

Binding agent: *Lignosulfonates* are very appealing industrial binding agents, both in terms of effectiveness and economic return. As adhesives in compressed materials, they perform as well, if not better, than petro-chemical alternatives. They are efficient as dust suppressants and structural reinforcement on unpaved roads. They can eliminate the environmental concern of airborne dust particles (especially in waterways), while stabilizing the road surface. In fact, lignosulfonates perform much better than calcium chloride in reducing dust, reducing the loss of road aggregate and lowering maintenance costs of unpaved roads (Road Saver Plus® 2004).

Approximately 5% of the dry lignin products made available for commercial use in Canada are used in animal pellets. By acting as a natural adhesive and helping maintain structural integrity, the addition of 2% lignin to animal pellets reduces the fine particles generated during handling and transportation by half (Lignin Institute 1992).

Dispersant: Lignosulfonate can also act as a dispersant, essentially preventing the clumping and settling of undissolved particles in suspensions, thus reducing the amount of water needed to use a product effectively. The use of lignosulfonates as concrete ad-mixers accounts for approximately 33% of the market share (Lignin Institute 1992). Adding lignin to concrete mixtures allows some control over the drying process. Depending on the ratio used, drying time can be increased or decreased, making the mixture more versatile.

Emulsifier: Lignosulfonates stabilize emulsions of immiscible liquids, such as oil and water, making them resistant to breaking.

Sequestrant: Lignosulfonates can tie up metal ions, preventing them from reacting with other compounds and becoming insoluble. Metal ions sequestered with lignosulfonates stay dissolved in solution, thus making them available to plants and preventing scaly mineral deposits in water systems.

4.1   Market Potential of Lignin

The demand for lignosulfonates will continue to grow as industries search for cheaper and more environmentally friendly sources of raw materials. The pulp and paper industry sells 1 million tonnes of lingo-sulfonates annually, making it the largest commercial source of lignin products. Tembec, one of the largest producers in the world, sells 570,000 tonnes of lignosulfonates annually (Arbo 2004).

Lignosulfonates can also be used as a base for polymers in the adhesive and plastics industry. In fact, they are a potential substitute for any products currently made from petrochemicals (Granit 2005).

The company Lenox Polymers Ltd., in Port Huron, Michigan, has developed and commercialized the technology to produce specialty resins from lignin as a replacement for various petrochemical polymers currently being used in the adhesive and plastics industry (Lenox Polymers Ltd. 2000). In Canada, Lenox is traded under the symbol LENP.

A German company called Tecnaro has been producing a lignin-based high quality thermoplastic (sold under the name Arboform) on a commercial scale since 2002, when they produced close to 300 tonnes. In 2002, the price of their product was US$5.50/kg (Schut 2002), but as demand for this type of plastic increases, and production increases proportionately, prices will decline, making the plastic a more feasible alternative. Tecnaro is currently using the lignin-based polymer to manufacture watches, figurines and flashlight handles. Other applications are automotive parts that look like wood but don't warp or crack.

5.   CONCLUSION

New advances in bioprocessing will make it possible for companies to use forest-based feedstocks in the production of biochemicals, and consumer demand for sustainable products will drive the market. In response, DuPont has pledged to derive 25% of its revenue from non-depletable resources by 2010, building on its 2002 record of obtaining 14% of its revenues from biobased products (DuPont 2003).

Royal Dutch Shell has predicted that biomass will supply 30% of worldwide needs for chemicals in the first half of the 21st century, while the US National Research Council predicts that over 90% of US organic chemicals will come from renewable resources by the end of the century. Other estimates from McKinsey & Associates and US Department of Energy foresee a rise from the 2003 global biochemical and bio-plastic market of US$60 billion to a 10% to 15% global market share in 2010 of US$140-210 billion. To support this trend, the US government has approved $250 million in research and development funds to accelerate the conversion of cellulosic materials to power, fuel and chemicals (Crawford 2001).

In Canada, the 2004 Innovation Roadmap on Bio-based Feedstocks, Fuels and Industrial Products has set targets for the chemicals and plastics sector to improve energy efficiency by 33%, representing a reduction of carbon emissions equal to seven megatons per year (Industry Canada 2004).

Many processing technologies are still at the developmental stage. In addition, the cost of transportation of raw materials is still a major concern. Money for research and development, government "Green Procurement Policies", tax incentives, and linkages and coordination between different industrial sectors will all be required before the biochemical industry can truly expand in Canada, with the forest sector as a major player.

Chapter 4

AGROFORESTRY

Highlights and Fast Facts

🌲	Agroforestry is being revisioned as a critical provider of goods for a sustainable bioeconomy.
🌲	Agroforestry systems include forest farming (sun and shade), silvopasture, timberbelts and integrated riparian (wetlands).
🌲	Bioproducts originate from organic matter, either newly grown or discarded, produced through wild harvesting ("wildcrafting"), agriculture (domestication) or agroforestry ecosystems.
🌲	Agroforestry involves the manipulation of natural ecosystems to promote or increase bioproduct species populations "in the wild".
🌲	To be labeled as an agroforest, ecosystem management must meet the criteria of being intentional, intensive, interactive and integrated.
🌲	Carbon sink potential in timberbelt agroforests is 8.5 tonnes per ha, leading to the potential for increasing numbers of tree plantations.

1.   AN INTRODUCTION TO AGROFORESTRY

Agroforesty, an integration of agriculture and forestry practices, was initially developed in tropical countries, driven by increasing human population density and deforestation mixed with extensive cattle farming. In temperate ecosystems such as Canada's, familiar agroforestry interventions are still in their infancy, although a strong case is now being made for the economic potential of cultivating plants in a forested area.

71

Bioproduct inventories originating from Canada's forests can be categorized into wild harvesting, domestic harvesting (agriculture), or agroforestry. Wild harvesting, also known as "wildcrafting", includes the gathering of naturally occurring plants from native forests (*e.g.*, lowbush blueberries (*Vaccinium angustifolium*) Wildcrafting is the extensive harvesting of wild plants in ways that will help increase the viability of a specific plant population. The terms wildcrafters, gatherers and harvesters often are used interchangeably (Letchworth 2001).

Intensive agricultural domestication of a species (*e.g.*, highbush blueberry (*Vaccinium corymbosum*) farms or cultivated ginseng) represents the transfer of bioproducts from their original forest home to the cultivated field. Agroforestry, on the other hand, encompasses both extensive and intensive manipulation of undomesticated, natural ecosystems to promote and/or increase the bioproduct species populations "in the wild".

The development of sustainable agriculture has helped to bring about a shift in perception, as agroforestry also presents the opportunity to develop new patterns in biomass production (Gordon and Newman 1997). As a system, it provides an ecologically-based approach to land management that contributes to biodiversity and ecosystem health, as well as long-term profitability within rural settings. While protecting biodiversity, reduced production costs through agroforestry play a critical role in business decision criteria where financial investments required are often modest. More challenging is the reorientation and training of various resource managers to appreciate the interdependence of their activities.

Agroforestry has similar characteristics regardless of the geographical location where it is practiced. It is basically an approach to land-use that incorporates trees into farming systems, and cultivation into forested areas. Within a global context, the Food and Agriculture Organization of the United Nations defines agroforestry as the art and science of sustainable land use that combines natural or planted trees and shrubs with crops and/or livestock on the same unit of land, in ways that increase and diversify farm and forest production while also conserving natural resources (Food and Agriculture Organization of the United Nations 2005).

At a national level, Natural Resources Canada defines agroforestry as the deliberate integration of trees and shrubs with non-woody crops and/or animals on the same land area for ecological and economic purposes (Small Woodlands Program of British Columbia 2001, Canadian Forest Service 2002). This chapter describes the five major agroforestry management systems emerging within temperate ecosystems (see Figure 4-1).

These agroforestry management systems include:

1. Forest Farming Agroforestry (FFA) Sun Systems (includes alley-cropping or intercropping).

2. FFA Shade Systems.
3. Silvopasture Agroforestry (SA).
4. Timberbelts Agroforestry (TA).
5. Integrated Riparian Agroforestry (IRA).

Shade Systems

Shade systems involve the intentional manipulation of trees, shrubs and ground vegetation to produce both timber and shade-requiring non-timber crops.

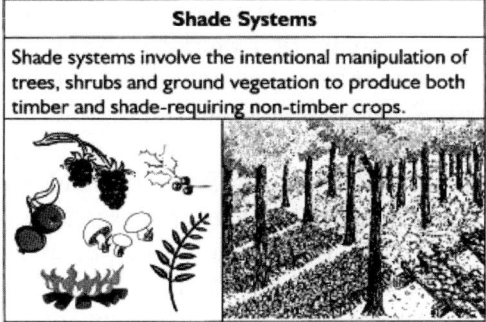

Silvopasture

Silvopasture systems combine trees with forage and livestock production.

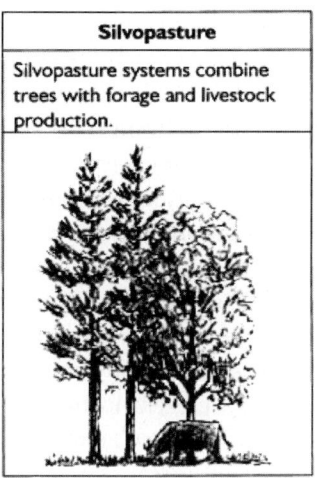

Integrated Riparian Management

Integrated riparian management is the management of areas bordering waterbodies and watercourses to enhance and protect aquatic resources while generating economic benefits.

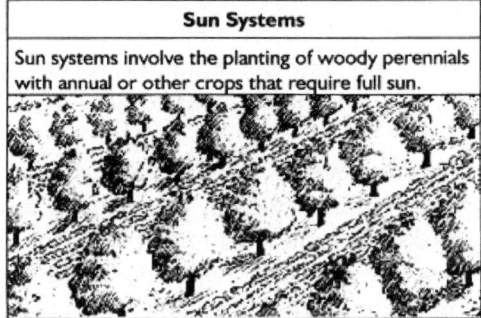

Timberbelts

Timberbelts consist of multiple rows of trees planted for both environmental protection and production of crops.

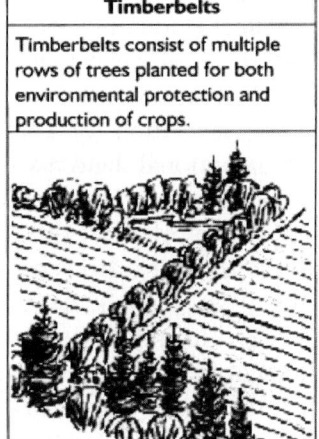

Sun Systems

Sun systems involve the planting of woody perennials with annual or other crops that require full sun.

Figure 4-1. Agroforestry.
Source: Natural Resources Canada, Canadian Forest Service.

From an ecological standpoint, agroforestry contributes to the conservation and protection of natural resources and to biodiversity. Whether extensive or intensive, these land management systems optimize benefits from ecological and economic interactions created through symbiotic combinations of trees, shrubs, crops (domestic and/or wild) and livestock. Agroforestry can also include short-rotation biomass plantations (Gordon and Newman 1997), which are cultivated stands of renewable, fast-growing "tree crops" for timber or biofuel/bioproduct purposes, as a way to preserve old growth forests or provide alternative energy, chemicals or materials. Overall, its method of intentional but sustainable combinations of organisms within a single system attempts to create harmony between environmental protection and human needs.

Renewed global interest and insights about stewardship of land and resources are changing perceptions about agroforestry concepts, providing a catalyst for expansion of agroforestry practices, bioproduct development and the valorization of rural environmental resources. New kinds of enterprises are thus increasing and diversifying the incomes and economy of our forests, agricultural lands, buffer zones and urban areas.

Indeed, interest in bioproducts from boreal and cold temperate forests is reflected through Canadian universities and colleges where the demand for education in agroforestry is increasing. As well, the University of Missouri was mandated to proceed with applications of agroforestry in North America, where the head office of the Association for Temperate Agroforestry has been since 1991.

There are four key criteria that must be satisfied in order to label the management of an ecosystem as an agroforestry practice. These characteristics identify agroforestry as intentional, intensive, interactive and integrated (Association for Temperate Agroforestry 2004). To meet the criteria of an intentional land use practice, combinations of trees, crops and/or animals must be managed as a whole unit, and not randomly occur as combinations of individual entities located within close proximity of one another, yet controlled separately.

Intensive management, be it concentrated or widespread, involves diverse, recurring silvicultural prescriptions that maintain productivity and ecological health (*i.e.*, cultivation, fertilization or inoculations of mushroom spores into logs). Agroforestry is considered interactive where the biological and physical connections between the individual components within the ecosystem are manipulated to enhance production of more than one harvestable commodity while providing conservation benefits (*i.e.*, controlling water pollution, promoting wildlife habitat).

Finally, within the ecosystem under management, individual components of trees, crops and/or animals must be structurally and functionally integrated into one single management unit, increasing productive capacity while at the same time balancing the resource conservation of the land.

1.1 Key Features of Agroforestry Systems

Agroforestry systems all have a common theme, which includes their deliberate design, with multiple goals that meet a variety of needs and that are subject to constant change (Small Woodlands Program of British Columbia 2001).

Practical experience instructs wildcrafters that plants are not found uniformly throughout the forests. Their production and abundance varies with a wide range of environmental and internal plant factors such as soil condition, climate, growing season, history of disturbance, vegetation type, plant age and condition (Duchesne et al. 2001). While choosing from a wide variety of silvicultural practices that are unique to each individual management system, the potential mixture of marketable crops and environmental benefits must be kept in mind (Small Woodlands Program of British Columbia 2001).

Agroforestry has the potential to reap great rewards (see Table 4-1), by multiplying existing plants, removing competing plants (while selling marketable materials), introducing plants and trees, or managing livestock.

Table 4-1. Bioproduct Potential from Agroforests.

Bioproducts Category	Examples
Wild Edible Food Products	Mushrooms, berries, herbs, vegetables and spices, honey, tree saps, tree nuts, wild rice, understorey plants, essential oils, seeds, teas, flavoring agents.
Materials and Manufacturing Products	Platform chemicals (polylactic, levulinic acid), bioplastics, silvichemicals (lignosulfates), essential oils.
Health and Personal Care Products	Pharmaceuticals and neutraceuticals, aromatherapy oils, cosmetics, herbal health products, fragrances.
Decorative and Aesthetic Products	Florals and greenery (salal), craft products, Christmas trees, native crafts, specialty wood products and carvings, cones.
Environmental Products	Biofuels, biopesticides, agrochemicals.
Landscape and Garden Products	Transplants (trees, shrubs, wildflowers, grasses), mulches, soil amendments.
Non-consumptive Bioproducts	Carbon credits, tourism and education, biodiversity conservation, recreation, water quality.

Adapted from: Mohammed 1999.

Unlike agricultural monocultures or conventional forestry timberstands, agroforestry systems are both diverse and dynamic, reflecting their constant change over time. This is because various life cycles within ecosystems are associated with natural or intentional forest disturbance. For example, small fruit-bearing species such as raspberry (*Rubus* spp.), mulberry (*Morus* spp.), saskatoon berry (*Amelanchier* spp.) and pin cherry (*Prunus pensylvanica*) are early colonizers after disturbance in forests; they thrive in open sunlight, and as such, may be included in management plans for FFA sun systems. However as this sun agroforest ecosystem matures, trees grow taller and plant competition increases for both light and nutrients, promoting the eventual displacement of these species from the ecosystem.

As a result of these changes, succession within ecosystems promotes the introduction and colonization of different harvestable species (Duchesne et al. 2001). Thus, what began as a FFA sun system has, over time, become a FFA shade system, supporting shade-tolerant species such as wild ginseng and clubmoss (*Lycopodium* spp.), which develop best in the understorey of older forests. As a result, agroforesters need to look ahead much further than is normally the case with monoculture systems.

1.2   Biology and Ecology as Mitigating Factors in Agroforest Planning

Understanding how plants grow and reproduce is central to the biology and ecology within agroforest ecosystems and subsequent bioproduct production. The huge number of bioproduct species available for harvest, coupled with the diverse habitats within which they are found, creates a knowledge gap regarding their biology and ecology as compared to that of forest tree species.

In order to incorporate bioproducts into conventional timber ecosystems, this challenge will need to be resolved. Information on plant physiology and morphology should lead to solutions regarding factors that control bioproduct establishment and distribution (Davidson-Hunt et al. 2001). The ecological study of bioproducts in agroforests should also focus on their temporal and spatial occurrence within ecosystems, drawing upon both traditional and scientific knowledge to create sustainable production.

1.3   Sustainability Within Agroforests

Most traditional (local) ecological-based knowledge and management systems emphasize ecosystem processes rather then production, and as a result, can support the diversity and sustainability within ecosystems that agroforests of the future should provide. Scientific principles of plant reproduction and growth also provide a guide for developing practices for sustainable use of the resource within agroforestry operations. According to Marles (2001), economic and cultural sustainability are considered essential for long-term productivity, consistent with local community needs and desires.

In designing agroforests to include the sustainable production of bioproducts, the concerns of indigenous peoples regarding commercialization of their traditional species must be met (Turner 2001). This not only includes consideration of indigenous intellectual property rights and cultural appropriation, but application of their philosophies using practical strategies of habitat enhancement and diversification through various time-honored management practices.

2. AGROFORESTRY SYSTEMS

As noted in Figure 4-1, there are five major agroforestry management systems existing within Canada's temperate ecosystems, with opportunities for a sixth system known as special applications agroforestry (SAA) existing within the framework of all agroforestry ecosystem practices. Each of the six systems is described in detail below.

2.1 Forest Farming Agroforestry - Sun Systems

Crops grown in FFA sun systems require full sunlight to mature, as with most berry crops. When these species are established or enhanced within existing forests that meet this criteria for light, the trees are found in random patterns and do not produce enough shade to qualify as a shade agroforest (Small Woodlands Program of British Columbia 2001, Canadian Forest Service 2002). This occurs in instances where stands of trees are too small, immature or spaced far apart, allowing planners to begin with a sun agroforest before transitioning to a shade system as the trees mature.

FFA sun systems where trees do not exist in a random or natural pattern are referred to as alleycropping or intercropping. The Association for Temperate Agroforestry describes this system as the cultivation of food, forage or specialty crops between rows of trees and as a larger version of intercropping carried out over longer periods of time (see Fig. 4-2).

Figure 4-2. FFA Sun System (Alley Cropping).
Source: US Department of Agriculture. National Agroforestry Center.

The greatest advantage of alleycropping or intercropping is that standard farm equipment may be used during a crop rotation. Tree and shrub crops do not necessarily have to grow in rows as pictured in Fig. 4-2, but may occur in blocks, clumps, grids or individual arrangements, as long as an open canopy exists to permit incoming light. In eastern Canada, many of these FFA sun systems provide income diversity by growing high value trees (*i.e.*, black walnut (*Juglans nigra*), nut or timber trees) along with other crops that have shorter rotations (*i.e.*, Christmas trees, forages and grains, horticultural plants like tomatoes and corn), to provide cash flow prior to their maturity.

When these agroforestry systems mature, they physically change and unless the trees are pruned or thinned, they cast more shade (see Fig. 4-3).

When shade levels increase, the introduction of shade-tolerant crops such as ginseng, mushrooms and floral greens may begin. Such a maturing agroforestry system can also be converted to silvopastural agroforestry (SA) if, for example, shade tolerant legumes are planted and livestock is allowed to forage. Since the trees are already mature, many of the problems associated with livestock damage to younger trees in SA are eliminated (Association for Temperate Agroforestry 2004a).

Importantly, slow-growing, high-value trees are not the only FFA sun system option and short-rotation trees, such as hybrid poplar, have proved profitable and viable when introduced to idle or marginal farmland (Association for Temperate Agroforestry 2004a). The significance of developing short-rotation plantation/agroforestry systems on marginal agricultural land for bioenergy, bioproducts and GHG reductions is currently emphasized by Masse (2005), who states that if 1.3 million hectares are established by 2025, 23 Mt of feedstock annually would provide 4.1% of the total energy consumption in Canada in addition to a reduction of GHG emissions by 30 megatonnes. Production of woody and herbaceous crops would occur through three production technologies: short-rotation intensive culture (primarily willows), afforestation plantations of fast-growing trees (mainly hybrid poplar) and agroforestry (FFA Sun System alley-cropping and riparian buffers).

Figure 4-3. FFA Sun System Alley Cropping (increasing shade).
Source: University of Missouri Center for Agroforestry.

2.2 Forest Farming Agroforestry - Shade Systems

Within forest farming agroforestry (FFA) shade systems, high-value understorey crops, either native or introduced, are intentionally grown under the protection of natural or planted forest canopies that have been modified in order to provide the correct shade levels (Fig. 4-4). Ginseng, pine mushrooms and other shade-loving species can be grown either in an intensive or extensive system, and their harvest will provide income during the time when high quality trees are being grown for wood products. It is important to note that wildcrafting of naturally occurring plants is not considered FFA.

Figure 4-4. FFA Shade System.
Source: US Department of Agriculture. National Agroforestry Center.

FFA shade systems are usually extensive, occur over a large area, incur lower labor costs and investment returns and promote the use of silvicultural techniques (*i.e.*, irrigation or controlling competing understorey vegetation) that improve the understorey species quality, quantity, diversity and sustainability. By comparison, an intensive system will involve a higher degree of labor investment, be more rigorously managed and usually occur on smaller areas, targeting higher value crops. The injecting of mushroom spores into logs under a forest canopy would be a prime example demonstrating intensive management.

Wills and Lipsey (1999) and Freeman (1997) studied a 5,988 ha parcel within the Nahatlatch Watershed of BC to determine a rough estimate of the comparative market value of pine mushrooms vs. timber values. Nahatlatch watershed consists mainly of open stands of Douglas fir (140-250 years old), plus hemlock, balsam (*Abies* spp.) and some additional minor species. Based on a 1998 six-month volume/price data analysis of this timber, and assuming that this land was clearcut now and once again after 120 years, the two-harvest value of the timber would be $426M (1998 dollars). Pine mushrooms wildcrafted at 1997 levels would generate an additional $73M over the same 120-year cycle for a combined total of $499M. Should Asian agroforestry practices, such as those practiced at the Kyoto Forest Experimental Station, be implemented to convert the Nahatlatch Watershed into a FFA shade system, high quality timber and pine mushroom harvests would generate $1.28 billion over the same time period.

2.3   Silvopasture Agroforestry

Silvopastural agroforestry (SA) involves the intentional combinations of trees, forage plants and livestock within an intensively managed system (see Fig. 4-5). The main categories of these combinations include forest grazing, where livestock exists within a forest, and intensive tree pastures, where trees exist within a pasture (Small Woodlands Program of British Columbia 2001). Nut crops, Christmas trees, maple products, fruit and berry crops, soft and hardwood lumber crops, forage, hay, meat and dairy can all be combined to generate and diversify income streams.

Figure 4-5. Silvopastural Agroforestry.
Source: US Department of Agriculture. National Agroforestry Center.

Where appropriate, forest grazing has the added benefit of livestock lowering fire hazards by removing groundcovers, which in turn compete for moisture with trees and other species (Association for Temperate Agroforestry 2004). Alternatively, intensive tree pastures involve the intro-duction of trees onto existing pastureland. Examples would include combinations of livestock with poplar and oak (*Quercus* spp.) plantings that could be managed for both short-term cash crops (*e.g.* poplar for firewood) with oak for longer rotation timber crops, while at the same time allowing for a continuous profitable livestock operation.

Within the tempered microclimate created under tree canopies under SA regimes, livestock benefit from reduced weather-related stress from wind and heat (through increased shade), along with higher forage quality and a longer growing season. Alternatively, tree crop yields will increase from the direct benefits from nitrogen-fixing forage crops, pasture fertilization and animal manure. Although livestock damage to trees, fencing costs and increased wild animal populations are a disadvantage, SA will succeed when management practices for trees, forage and livestock are compatible.

2.4   Timberbelt Agroforestry

Timberbelt agroforestry (TA) consists of windbreaks and fenceline planting of trees and shrubs with the intent of providing environmental benefits (wind protection, soil conservation and wildlife habitat), in conjunction with increasing economic benefits through the added production of tree crops and other bioproducts.

Windbreaks typically consist of linear plantings of trees that are perpendicular to the prevailing wind direction. In the past, single rows of trees were the norm (monocultures); however, multiple row plantings of different species have distinct advantages and are more accurately named timberbelts (see Fig. 4-6). Multiple row plantings provide better wind protection and allow the manager to harvest trees without complete removal of the windbreak. Prospects for wildlife enhancement are also improved by providing ecological niches within which they may live and corridors through which they may pass.

Under the canopy within multiple row windbreaks, FFA shade species may be introduced to diversify income streams. Environmental benefits will apply to both livestock and people through the conservation of soil and water, by managing snow dispersal to control drifting to keep roads clean or spreading snow evenly across a field to capture water (Small Woodlands Program of British Columbia 2001).

Field windbreaks also provide a variety of protection for wind-sensitive crops, boosting crop yield and quality, in addition to reducing erosion, and increasing bee pollination and pesticide effectiveness. Other benefits include the use of windbreaks as buffers between urban, rural and forestry areas to control dust and odors originating from livestock and machinery, or to reduce heating costs in homes.

Figure 4-6. Timberbelt Agroforestry.
Source: US Department of Agriculture. National Agroforestry Center.

Fenceline plantings quite often consist of a single row of trees and offer great potential to utilize unused land. According to the *Guide to Agroforestry in British Columbia* (Small Woodlands Program of British Columbia 2001), in lower mainland BC alone, there is potential to manage up to 50,000 ha of fencelines for trees and shrubs. These sorts of ideas are well described by the Prairie Farm Rehabilitation Administration Shelterbelt Centre in Saskatchewan (Agriculture and Agrifood Canada 2004). To deal with the dust bowl of the 1930s depression era, the Prairie Farm Rehabilitation Administration began by providing tree seedlings at low cost for the establishment of windbreaks. Today, many new initiatives are being researched and developed through the Prairie Farm Rehabilitation Administration, with values ranging within TA systems only limited by the rate at which they may be established.

The Association for Temperate Agroforestry (2004) estimates that the carbon sink stored within the trees by age 20 within each ha of planted TA equals 8.5 metric tonnes. In the Climate Change Action Plan, the Government of Canada has acknowledged that enhancing existing timberbelts reduces greenhouse gases (GHGs) under the Kyoto Protocol (Government of Canada 2000) and they have designated the planting of TA as a "best management practice" for sequestration of atmospheric carbon (Agriculture and Agrifood Canada 2004). Combining carbon sequestration with growing trees on plantations for renewable bioenergy to offset fossil fuel usage (and potentially generate carbon credits – see Chapter 12) could lead to even greater reduction of GHGs through agroforestry.

2.5   Integrated Riparian Agroforests

A riparian zone is defined as land immediately bordering a watercourse or water body including rivers, creeks, drainage ditches, lakes, sloughs, wetlands, canals and springs. Vegetation found within typical prairie riparian zones often include sedges, grasses and shrubs such as willows and redosier dogwood. Alternatively, integrated riparian agroforests (IRA) often will have greater biodiversity and may include trees, shrubs, sedges and grasses along with a variety of FFA shade and sun species. They also may be considered a buffer or zone of transition between a water body and upland vegetation, cropland or pasture. IRA involves the management of these riparian zones to improve and protect aquatic resources while generating economic benefits (see Fig. 4-7).

Figure 4-7. Integrated Riparian Agroforestry.
Source: US Department of Agriculture. National Agroforestry Center.

Ecological benefits of creating and maintaining a healthy riparian forest buffer include improvements to water quality, maintenance of productive aquatic environments, filtration of sediments and agricultural pollutants, streambank stability, erosion and flood control, enhancement of wildlife habitat, protection of soil productivity and increased biodiversity.

According to Agriculture and Agri-Food Canada, the biodiversity found within riparian ecosystems is quite possibly the most productive and richest in the world (Agriculture and Agri-Food Canada 2000). Throughout the Canadian prairies, most species of wildlife will spend at least some part of their life cycle within these ecosystems or use them as travel corridors from one area to another.

Economic benefits derived from IRA also are diverse and range from aesthetics and recreational values (*e.g.*, sport fishing) to a diverse list of bioproducts which may be grown in IRA, including botanical, medicinal, floral and edible products, along with traditional products such as Christmas trees, timber production and wildlife recreation (Association for Temperate Agroforestry 2004).

If an intact IRA is already vegetated, then the management plan will be similar to that for a FFA shade system, with special attention paid to existing regulations regarding these sensitive areas. If the riparian area is degraded, the primary objective will be to restore ecological function (Small Woodlands Program of British Columbia 2001). However, the possibility always exists to establish a tailored agroforestry system that will provide both economic and ecological benefits.

2.6　　Special Applications Agroforestry

Opportunities for special applications agroforestry (SAA) exist within the framework of all previously discussed agroforestry ecosystem practices (*e.g.*, visual screening, noise and odor control), where enhancements can produce unique commodities. An example of SAA would be the planting of short-rotation woody crops into timberbelts to sequester carbon, enhance wildlife and improve water quality while at the same time providing short-term income. The United States Department of Agriculture's National Agroforestry Center (2004) provides some other examples:

1. Wastewater Management – Using hybrid poplar and willow to absorb agricultural, industrial and community wastewater and produce short-rotation wood fiber.
2. Waterbreaks – Planting trees on floodplains to reduce flooding impacts by moderating water flows in a fashion similar to the way timberbelts control wind speeds.
3. Biotechnical Streambank Protection – Involves the use of plants, grasses and woody vegetation to reinforce soil, stabilize slopes and reduce erosion.

4. Working Trees for Communities – Involves a multitude of urban and community agroforestry applications to improve soil, water and air quality, provide wildlife habitat and travel corridors, recreational opportunities (green space), visual and noise screens and carbon storage that support healthy communities.

3.   ADVANTAGES AND DISADVANTAGES OF AGROFORESTRY PRACTICES

Agroforestry practices have unmistakable advantages in terms of soil conservation, coupled with reduced rates of erosion, promoting outstanding results for water conservation and aquatic environmental quality. The use of pesticides also is dramatically reduced since plants are not usually grown within monocultures and can take advantage of their natural environment to repel pests.

For example, the intensive production of American ginseng (*Panax quinquefolius*) in an open field requires costly infrastructure to create shade and, due to plant density, the frequent use of pesticides and fungicides, resulting in a loss of ginseng quality (Nadeau et al. 1999). Growing ginseng in a FFA shade system, however, eliminates the need for shade structures and drastically reduces pesticide and fungicide requirements.

Agroforestry practices also can help relieve the pressures on wild plant harvesting where well-managed controls on cultivated plants can restrain erratic harvesting and protect natural stocks. This would help to alleviate current concerns that wildcrafting (particularly of rare native species) may represent a risk for biodiversity because of overexploitation. It can also contribute to the preservation of old growth forests.

Primary factors that currently limit agroforestry development and scope of application have been identified by Olivier (2001) as follows:

1. Labor requirements are higher, and in most cases, mechanization of the operation is still required.
2. Further research is essential to have optimal production in agroforests.
3. There is a need to develop more agroforestry techniques.
4. Marketing of bioproducts is not yet fully established.
5. There may be a need for legislation or other policy incentives.

The actual economic value derived within present day agroforestry systems also is not well defined. However, the current systems of intensive agricultural and forestry practices lack diversity and create income streams that are but a fraction of their future potential. With the help of research and technological innovation, agroforestry techniques can provide supplementary income to producers and managers and create new avenues for bioproducts.

From another, narrower, point of view, fear of an increase in productivity disrupting supply and demand economics could lead to lower wholesale commodity prices. For example, in April 2001, the price of many medicinal plants (*Echinacae* spp., *Ginkgo biloba*, St. John's wart (*Hypericum perforatum*), etc) dropped 50% to 75%. According to a Richters Herbs Commentary in 2002, without industry-wide standards combined with general expanded interest for bioproducts at the retail level, it may be in the producer's best interest to cultivate a limited volume of plants to maintain the price while reducing inventories.

To increase the economic and social contribution of agroforestry, various constraints must be addressed, including: 1) establishing a foothold on a land base that is non-competitive with the current forest and agriculture industries; 2) creating products that are cost-competitive with the traditional commodities agroforestry seeks to displace; 3) promoting value-added opportunities for which markets must be developed; and 4) establishing proof-of-concept research that will support the requirements of the investment/entrepreneurial sectors.

Despite these daunting challenges, there are many exciting opportunities for the agroforestry sector to re-position itself as a key provider of the feedstocks necessary for the development of consumer goods in the emerging bioeconomy. A re-visioned Canadian agroforestry sector can rely on a sustainable growth based on a number of competitive advantages:

1. Rural, remote and First Nation communities are in search of novel initiatives to promote economic and social development, and, as such, are becoming investors in innovation as well as a secured market.
2. Agroforesters potentially have access to an abundance of marginal agricultural lands, power line routes and non-productive forests.
3. Nationally, there are increasing economic and political incentives to promote reliance on alternative, renewable forms of energy.
4. Carbon sequestration and bioenergy produced from dedicated tree plantations or from waste streams is carbon neutral, and can generate carbon credits by offsetting the use of fossil fuels.
5. The financial sector is aware of the impact of energy pricing of business and sees renewable fuels, chemicals and materials from agroforestry as a potentially significant commercial opportunity.

4.   CONCLUSION

The growing fragility worldwide of both natural ecosystems and agricultural systems is generally known. In the traditional agricultural setting, repetitive working of the soil takes place each year at the beginning of the plant growing process and again after the harvest, exposing bare soil to wind and water erosion. The need for plowing, harrowing, fertilization and pesticide applications requires many passes over the field with heavy, high-consumption machinery. Topsoil erosion, combined with degradation of aquatic and aquifer systems, downward leaching of fertilizers and agricultural pesticides, as well as increasing energy needs, are serious challenges, partly resulting from these intensive production methods.

In contrast, most agroforestry systems of cultivation in forested ecosystems are established for longer periods of time and are less intense and invasive. Changing the way resources are viewed, to promote the sustainable application of agroforestry practices, has the potential to enhance forest biodiversity and create more varied and innovative agricultural systems in Canada.

Finally, by supporting the agroforestry sector through the development of market-ready bioproducts, agroforestry can be repositioned as a critical provider of goods for the bioeconomy. Increased markets, profits, tax advantages, environmental enhancement, efficient land use and sustainability for future generations are but a few of the benefits that can be derived from a new vision of agroforestry in Canada.

Chapter 5

FOODS FROM THE FOREST

Highlights and Fast Facts

- 🌲 The estimated economic output of forest based foods (FBF) in the Canadian economy ranges from $725 million to $1.33 billion.
- 🌲 Future economic potential of FBF is projected between $2 and $7.4 billion per year.
- 🌲 With 8% to 14% annual economic growth, functional foods represent a $1.25 billion industry while reducing the risk of chronic diseases and providing consumers with many other benefits.
- 🌲 Canada is a world leader in organically certified foods, with 20% annual economic growth and $3.1 billion in estimated sales.
- 🌲 Canada produces 85% of the world's maple syrup and 60% of the lowbush blueberries on the market.
- 🌲 One single pine (matsutake) mushroom sold in Japan may retail for up to US$150.
- 🌲 If wild ginseng were domesticated and grown in agroforests, its retail value would increase 10-fold.
- 🌲 As an indigenous resource and unique Canadian product, wild rice needs protection from "genetic pollution".

Although the greatest capacity for Canadian agricultural production has its genetic origins in southern plains and grassland ecosystems, Canada's forests hold immense promise for the development of forest foods and subsequent northern community prosperity. In the context of forest-based bioproducts generally, forest-based foods (FBF) include not only non-timber forest products but also any plant life that has its genetic origins in a woodland or forest ecosystem. All plants dependent on any temporal phase of a forest ecosystem or on any stage of forest succession can, therefore, be included within this category.

The growing conditions of common lowbush blueberries (*Vaccinium angustifolium*), for example, are dependent on the cycle of forest clearance and renewal (*i.e.*, succession), and are also adapted to forest clearance, typically a post-harvest or post-fire situation. The fact that highbush blueberries are now commonly grown in permanently non-forested fields should not detract from their origins in the forest.

Similarly, honey is in its origins a forest product and although virtually all marketed honey production is now associated with agriculture, forage among flowering trees remains an important source of nectar for bees. As with blueberries, the transition of honey from a wild FBF to a domesticated commodity provides valuable lessons regarding possibilities for transitioning other FBF into similar production systems or into agroforests which can enhance productivity and diversify market potential.

Future FBF will have especially significant markets if they are amenable to development within agroforestry systems or in agricultural contexts as domesticated crops. Through such diversification, a range of FBF could make the transition from obscure specialty items to products with considerable consumer appeal, warranting greater investment in research and economic infrastructure supporting FBF in Canada.

1.   CURRENT SCALE OF PRODUCTION OF FOREST-RELATED FOOD BIOPRODUCTS

FBF crops such as honey, maple syrup, berries, wild mushrooms, understorey plants and wild rice (*Zizania* spp.) already make a significant contribution of over $725M to the Canadian economy (Table 5-1), with future economic potential of greater than $2 billion. The potential market value of FBF could exceed $7.4 billion if we add the value of agricultural crops that had their origins in wooded ecosystems (*e.g.*, tree fruits, nuts, cultivated berries, cultivated mushrooms, ginger) in addition to functional and organic foods.

Regional production and economic data are as yet insufficiently developed regarding the role and economic value of FBF as a source of ideas for new value-added food products, molecular material for functional foods and nutraceuticals, genetic information for novel and existing domesticated food species, secondary (non-food) goods, and services and habitat for animal species. Although comprehensive species lists of FBF have been compiled (Langlais 2003; Mohammed 1999; Turner & Szczawinski 1988;

Wills & Lipsey 1999), the future potential of these products could be far more significant than presently realized.

Table 5-1. Estimated Current Output of FBF in the Canadian Economy[1].

FBF Commodity	Output in Tonnes or Litres $'000	Current Economic Value $'000	Additional / Future Economic Potential[2]	
			Secondary Products and/or Additional Potential[3]	Value $'000
Honey	37,072	160,805	*Pollination[4]*	*1,000,000*
			Wax, royal jelly, propolis[5]	804,025
Tree Saps	34,761	163,968	Additional maple potential	164,000
			Birch syrup products (Alta)	31,200
Berries	149,373	278,654	Specialty niches: "native fruits"	26,000
Wild Mushrooms	1.1397	43,000	Edible export potential	115,000
			Medicinal and Nutraceutical potential[6]	>115,000
			FFA Matsutake Potential[7]	1.183/ha/yr
Understorey Plants	2.30	75,321	FFA ginseng 10x field cultivated roots (Agriculture and Agri-Food Canada 2000a)	753,210
Wild Rice	1013.206	3,492	Qualifies for OCIA[8] Enormous potential	3,492
Functional Food		*800,000*	8% to 14% growth / year	*1,250,000*
Organic Food		*500,000*	20% growth / year	*3,100,000*
Other FBB		31,088	*Herbal Teas, Tree Nuts, Essential Oils*	66,412
Total		725,240	Total	2,011,927
		1,331,088		*5,416,412*

[1]Calculations are found in subsequent sections of this and other chapters.
[2]In some instances the economic potential currently exists; however, "official" statistical values have not been compiled.
[3]*Where FBF potential cannot be disaggregated from conventional agriculture, values are italicized and not included in the total.*
[4]Agriculture and Agri-Food Canada 2003. Value of pollination.
[5]Mitchell M. and Associates 1997. Honey by-products have a wholesale price at minimum five times the cost of base honey ingredients.
[6]Wills and Lipsey 1999. Additional FBF mushroom potential for medicinals and nutraceuticals > $115 million.
[7]Forest Farming Agroforestry (FFA) shade system potential (see Nahatlatch Watershed Study,
Chapter 4). Not included in total.
[8]Organic Crop Improvement Association International 2005.

1.1   Challenges with Data Collection

Given that sales data for most FBF discussed here are not monitored very well at the national level, a report on their economic potential within Canada should be viewed as a work in progress. The task of calculating the contribution of FBF to the Canadian economy is beset by a number of problems; therefore, it is worth briefly mentioning some of these difficulties at the outset.

To begin, nutritional and pharmacological values realized from FBF are not considered. Each of the food products reviewed likely has a host of medicinal properties making the distinction between food and health products somewhat arbitrary. The economic benefits of FBF in terms of health industry revenues, and secondary benefits of reduced government health care spending, are not yet calculated. Equally important, there is no standard method of data collection or a way to directly determine the value of FBF as sources of product innovation or genetic and molecular material.

Although Table 5-1 suggests an order of magnitude of FBF contribution to the national economy, the most significant immediate impact of further development would be realized at the local and regional level. The value of FBF to these economies is often more significant than national figures suggest. Local values can be calculated through intensive survey efforts for specific communities and regions, as has been undertaken through harvest studies with First Nations (Boxall et al. 2002). Opportunities exist for the development of FBF species to move toward full market potential with their own unique set of locally-based social and ecological constraints.

2.   MAKING THE TRANSITION TO GREATER MARKET POTENTIAL

Table 5-2 summarizes how FBF currently grow within three regimes: 1.) as wild foods in the forest harvested by wildcrafters; 2.) as wild foods in the forest with some degree of land management (agroforestry); and 3.) as wild foods which have been domesticated into agricultural settings.

As one moves across the proposed management continuum from wildcrafting through agroforesty to domestication (Table 5-3), producers can realize greater economic rewards, but generally at the expense of greater investment in research and infrastructure from both private and government sources. Domestication of FBF for agricultural production likely will be beneficial in the long run for larger southern Canadian producers, with established agricultural mechanisms and nearby markets. In order to benefit northern forest-resource communities, it will be necessary to implement various agroforestry management techniques in order to enhance yields of many other FBF now harvested or wildcrafted in their natural ecosystems.

Table 5-2. Current Production Regimes for Selected FBF.

Domestication (Agriculture)	Agroforestry[1] (Wild)	Wildcrafting (Wild Harvesting)
Honey		
	Maple Syrup	Maple Syrup
	Birch Syrup	Birch Syrup
High-bush blueberries	Low-bush Blueberry	Low-bush Blueberry
Cultivated strawberries Cultivated Raspberries Cultivated Currants	Cranberries Saskatoon berries	Wild Strawberries Wild Raspberries Wild Currants Highbush Cranberries Saskatoon berries Lingon berries
Shiitake Mushrooms	Matsutake	Matsutake Morels Boletes Chanterelles Lobsters
Cultivated Leeks Cultivated Ginger Cultivated Ginseng	Ginseng	Wild Leeks Wild Ginger Wild Ginseng Fiddleheads
Wild Rice ("paddy - grown", US)	Wild Rice ("Canadian Lake")	Wild Rice

[1]Refer to Chapter 4 (Agroforestry) for examples of this land management.

At the opposite end of this spectrum are those specialty FBF that are self-reproducing within a forest ecosystem, with little or no human intervention, but which cannot make the transition from wild harvesting to domestication. These specialized products, often targeted to "high-end consumers", focus on organic, bioregional, wild and "slow" foods. This market is actually preserved by the inability to provide alternate growing conditions outside of indigenous forests, ensuring that premiums will continue to be paid for these specialty products.

Table 5-3. Management Regime Continuum.

Criteria	Domestication (Agriculture)	Agroforestry	Wildcrafting (Wild Harvesting)
Species Objective	Monocultures. Improved cultivars.	Increased biodiversity. Organic certification.	Informal or none. Preservation.
Ecological Consideration	Farm-based agroecosystems. Consistent sustainable productivity.	Modified ecosystems. Consistent sustainable productivity.	Forest (natural) ecosystems. Highly variable productivity.
Economic Consideration	Bulk sales (economies of scale). Commodity markets (low prices). Lower profit margins. Centralized market planning.	Subsistence/bulk sales. Specialized formal markets. Medium profit margins. Increased market planning.	Subsistence sales. Specialized informal markets. Higher profit margins. Little or no market planning.
Infrastructure	Most infrastructure. Full mechanization. Private property.	Less infrastructure. Some mechanization. Crown/private land.	Little infrastructure. Labor intensive. Usually crown land.
Scale/Range	Regional/National. Smaller areas. Accessible.	Regional. Medium/large areas. Remote (especially if organic) or accessible.	Local/Community. Often larger areas. Usually remote.
Immediate Beneficiary	Agricultural and agri-food industries.	Provinces, forest managers.	Individual households and communities.

It would nevertheless be valuable to understand the growing conditions of these lesser-known FBF, as well as who is making use of them and how they might be brought to a wider market. Detailed biophysical data are required to understand what species are most suitable to transitioning out of the forest through the enhancement of growing conditions (agroforestry) or preservation of FBF-rich ecosystems.

At present, crown forest management is primarily conducted through forest industry managers, which generally means timber companies that hold long-term forest licenses. Care needs to be taken to avoid simply folding the benefits of FBF and other bioproduct development into existing management institutions so that long-term license holders become the major beneficiaries.

The development of FBF projects will be most successful when they expand the base of stakeholders interested in sustainable forest management that generates the greatest benefits for forest communities.

With all these factors in mind, forest management planning can take better account of the non-timber values in the forest, stimulating a greater range of benefits for a broader range of stakeholders, particularly in Aboriginal communities where agricultural production is very limited. Consideration, therefore, needs to be given to the development of appropriate policy instruments and incentives for FBF development.

2.1   Wildcrafting

As defined in Chapter 4, wildcrafting is the gathering of natually occuring plant material from its native or "wild" environment. However, with unregulated harvests and few exhaustive studies or harvest guidelines, negative impacts from wildcrafting forest plants have emerged. A historical example demonstrates how the demand for wild ginseng almost wiped out this species at the beginning of the 1800s, and only through domestication did ginseng supply meet consumer demand.

At present, over-harvesting of ground hemlock or *Taxus* has, in many parts of the world, led to the decimation of wild stands. To halt this process, research guidelines for harvesting ground hemlock were jointly developed in 2002 by Natural Resources Canada and the Prince Edward Island Department of Environment, Energy and Forestry, and focus mainly on pruning techniques based on the species' capacity for regeneration at a rate equivalent to harvest levels. Canadian wildcrafters who voluntarily follow these guidelines support sustainable management and support a viable long-term industry. In the United States, the Food and Drug Administration (FDA) is now enforcing recognition of sustainable harvest practices. The FDA requires that plant-derived products originate from sustainable managed sources confirmed with a harvest verification system requiring that all wildcrafted *Taxus* be inspected by a "third party" to verify that proper harvest methods have been used (Prince Edward Island Department of Environment, Energy and Forestry 2002).

Figure 5-1. Ground Hemlock (*Taxus*).

Source: M. Laporte.

2.2   Domestication (Agriculture)

In many instances, large commercial buyers have demonstrated that the supply of an NTFP often cannot meet the international demand for such a product which places the future of some of these species in question (Duchesne et al. 2001a). Commercial exploitation combined with the threat of habitat conversion, along with the potential for conflict with other forest users, will create new pressures on individual species and ecosystems. As a result, new strategies for the cultivation of these plants must be investigated to create a sustainable resource and avoid depleting forest biodiversity. Future research should concentrate on domestication, "semi-domestication" (agroforestry) or restoration of NTFP habitat for wildcrafting. Although many NTFPs are sustainable within forest communities, some are not, and those already, or will eventually, require domestication (see Table 5-2).

To attain successful domestication of a species, complete understanding of its genetics, biology and ecology will be needed to produce a product that is as attractive as that grown naturally (Duchesne et al. 2001a). This understanding can be acquired through traditional knowledge, as well as through conventional scientific research. This approach has been successfully demonstrated through the domestication of ginseng and highbush blueberries resulting largely from our understanding of their ecological requirements. Maximizing harvest, avoiding further depletion of natural stocks, stabilizing supplies and increasing economic returns will help these NTFPs make the transition to a greater market potential.

2.3   Organic Certification

Recent public attitudes and consumer trends have shown that the "functional" or "smart" aspects of foods can be further enhanced through organic certification. The farm gate value (FGV) of Canadian-grown organic foods has been steadily increasing by 20% per year over the past ten years. In 2000, the FGV for this industry equaled $500 million, and growth in this sector is expected to increase to $3.1 billion by 2005 (Agriculture and Agri-Food Canada 2003a).

Responding to this increased interest in organic products, the National Standard System for Organic Production in Canada, established to develop, promote and implement voluntary standards, was published by the Canadian General Standards Board (1999), in association with the Standards Council of Canada (Agriculture and Agri-Food Canada 2003b).

The National Standard System for organic agriculture was adopted in Canada in 2003, around the same time as the implementation of the United States Department of Agriculture's National Organic Program. If strict standards of organic production are met, products such as "Canadian Lake Wild Rice" can be certified and identified with labels endorsed by the International Organic Crop Improvement Association (Organic Crop Improvement Association International 2005). Section 8.6 of the Canadian General Standards Board guidelines describe what is required for wild and natural products to qualify for a "certified organic" label:

8.6.1   A detailed outline of the harvest area and 36-month history of compliance including a description of harvest methods used and proposed measures for the protection of wild species.

8.6.2   Wild natural products may only be certified organic if harvested from relatively undisturbed environments with harvesting not negatively impacting the ecosystem.

8.6.3   Wild and natural products must come from clearly delineated production zones subject to routine inspection.

8.6.4   The production zone requires clearly defined buffer zones.

8.6.5   The manager of the harvest must maintain an audit trail.

(Canadian General Standards Board 1999)

As these guidelines demonstrate, policy instruments play an important role in maintaining accuracy of labeling and promoting consumer confidence for wild and natural products, in addition to supporting sustainable management and species viability. Adhering to the above criteria will involve increased organic production expenses but should be offset by price premiums paid by consumers for the "certified organic" label.

3.   EXAMPLES OF CANADIAN FBF WITH SIGNIFICANT MARKET IMPACT

As there are innumerable FBF species in Canada, only the current production and future development potential of the most visible and viable food products will be examined in the remainder of this chapter.

3.1   Honey

Unlike many FBF, domesticated honey and honey by-products (pollination, wax, royal jelly, propolis) are a well established industry within Canada, with production and export data readily available from a variety of sources. Although accounting for only 3% of audited global honey production, Canada ranks as the ninth largest honey producer in the world, and is recognized for producing the highest quality of prized mild white honey, renowned for its taste (Agriculture and Agri-Food Canada 2003). China is the undisputed leader of world honey production by volume, while the United States and Argentina are the second and third largest producers, respectively (Agriculture and Agri-Food Canada 2003). Nevertheless, the average of all Canadian honey yields remains roughly twice the world average (Agriculture and Agri-Food Canada 2003).

The honey industry is intimately tied to agriculture where honey bees play a critical role in pollination, accounting for an estimated 80% of insect pollination of crops (Hester 2004). Not surprisingly, the prairie provinces produce 75% to 80% of Canadian honey (Table 5-4), where beekeepers take advantage of long summer days with fewer clouds, allowing greater time for bees to forage among the vast acreages of canola (*Brassica* spp.), alfalfa (*Medicago* spp.), clover (*Dalea* spp.) and borage. Bees require nectar and pollen over a long period of time to produce large quantities of honey. A succession of flowering periods is best for production where early season sources of nectar are available including cranberry, maple, willows, alder (*Alnus* spp.), pin cherry and dandelion (*Taraxacum* spp.) (Mitchell M. and Associates 1997). Where rural agricultural fields in the north are absent, bees utilize the many natural forest stands available in the vicinity of the hive (Mohammed 1999).

As Canadian's current annual per capita consumption is well below national production levels, it would make sense for future marketing to focus on exports of bulk honey and value added production. However, despite its strong production capacity, Canada imported 8,118 metric tonnes of honey in 2002, with a value of $23.242M (Agriculture and Agri-Food Canada 2003). Since honey production can be a very lucrative business with average profit margins estimated at 125% (Mitchell M. and Associates 1997), this represents a substantial potential market share for Canadian producers to fill. Recent Canadian prohibitions on Chinese honey imports (due to elevated antibiotic residues) and US (anti-dumping) trade restrictions action against Argentina and China (Agriculture and Agri-Food Canada 2003) are catalysts for increasing domestic production. Also, opportunities to expand production could be generated by the growing demand for natural and nutritious sweeteners as well as for bee by-products.

Table 5-4. Honey Production and Export Statistics in Canada.

Province	Beekeepers[1] (number)	Colonies[1] (number)	Production[1] (tonnes)	Value $'000	
				FarmGate[1]	Exports[2]
P.E.I.	55	2005	52	220	123
Nova Scotia	418	20,500	342	1,775	414
New Brunswick	230	5,400	145	595	188
Quebec	220	31,710	1,683	6,720	9,464
Ontario	3,000	75,000	4,824	20,915	5,216
Manitoba	800	87,000	6,511	28,710	33,011
Saskatchewan	1,325	100,000	8,618	36,100	23,506
Alberta	700	227,000	13,488	58,320	16,567
British Columbia	2,210	39,870	1,408	7,450	1,223
Canada - 2002	8,958	588,485	37,072	160,805	89,712
Canada 1998-2002[1]	9,601	588,434	37,513	97,438	42,663

[1]Statistics Canada (2003a). Five year averages: 1998 -2002. Does not include Newfoundland.
[2]Industry Canada (2005) - HS Codes.

These secondary products (wax, propolis, bee pollen, royal jelly and bee venom) are used in a variety of consumer goods including teas, lotions, salves, crafts, candles, waterproofing, polishes, confectionaries and other edible products. Forest-dependent Canadian beekeepers are potential major suppliers of high quality propolis, a brownish resinous material of waxy consistency collected by bees from the buds of trees and used as a cement. A high proportion of propolis is formed from tree resins, with poplar and alder being prime sources. It is also increasingly recognized for its nutritional properties, driving up price increases two to three times between 1990 and 2000 (Canadian Honey Council 2001). For all of these secondary bee by-products, wholesale prices are, on average, at least five times that of the base "honey" ingredient, yet a Canadian FGV for them has not been properly estimated.

Future developments appear dependent on resolving pest problems, such as the introduction of Africanized (killer) honey bees, American foulbrood, (a small hive beetle), and the tracheal and varroa mites. These pests have substantially reduced honey production in Europe and threaten the beekeeping industry in North America. Detection of the varroa mite led to a Canadian prohibition, since 1987, on live bee imports from the United States (Agriculture and Agri-Food Canada 2003a). Chemicals that could control the varroa mite are not used due to the fact that they are toxic to both bees and humans, can contaminate both honey and wax, and varroa mites are quick to develop resistance to treatment (Rickert 2004). Plus, with increasing domestic and international demand for organic and antibiotic-free honey, use of insecticides to overcome pest problems will only hinder Canadian exports.

Of course this growing demand for natural honey, coupled with the advance of debilitating pests in centers of agricultural production, can also be seen as real opportunity for small producers to take advantage of forest and meadow ecosystems in areas far removed from agricultural systems. Although production from wild sources entails lower yields per colony, beekeeping can provide a useful subsidiary income among forest communities. In such cases, sales should emphasize "value added" and organic products, following the example of the maple industry.

3.2   Tree Saps

Although trees that produce maple syrup have broad geographic distribution, only in northeastern North America do the warm spring days followed by below-freezing nights stimulate them to produce commercial quantities of sap. Tree saps are similar to slightly sweet water. Higher sugar content in the sap of sugar maple (*Acer saccharum*) ensures this tree's number one position for syrup production in Canada; however, both the red maple (*Acer rubrum*) and the sugar maple can be used. During each warm spring day, healthy maples that are at least 40 years old may produce up to five litres of sap and, if properly managed, the trees' productivity will continue for over a century. From each season lasting up to six weeks, one tree will yield 35-50 litres of sap which, when "boiled down", will produce 1.5 litres of syrup (Agriculture and Agri-Food Canada 2003c).

Canada's share of world maple syrup production rose from 62% in 1984 (Fédération des producteurs acéricoles du Québec 2004) to 85% in 2002 (Agriculture and Agri-Food Canada 2004a), with the eastern US accounting for the remainder. Canadian exports 81% of its syrup to the US, with Western Europe and Japan importing the rest (Agriculture and Agri-Food Canada 2004a). Maple product value in Canada is currently estimated to be at 50% of its potential (Agriculture and Agri-Food Canada 2001), with 93% of production originating in Quebec (Agriculture and Agri-Food Canada 2003c). In 2002, production was 34,761 metric tonnes with a value of $164M (Table 5-5). Exports in 2002 equaled $147M, representing a huge increase as compared to $42M only a decade earlier.

Table 5-5. Maple Products Production, Value and Exports in 2002

Source	Production[1]		Exports	
	Tonnes	Value $000	Tonnes[1]	Value $000[2]
Quebec	32,495	147,056	23,283	124,991
Ontario	1,376	11,063	4,018	15,964
P.E.I.				3,507
B.C.				1,703
N.B.	884	5,849		1,078
Others				92
Canada	34,761	163,968	28,685	147,335
US	6,965	38,389		

[1]Agriculture and Agri-Food Canada (2003c).
[2]Industry Canada (2005) - HS Trade Data.

Increases in production can largely be attributed to an increasing number of taps rather than numbers of farms, as the industry has been experiencing a progressive concentration of production among larger operations. Between 1981 and 2001, the number of taps in Canada rose by 99% during a period when the number of producers (farms) fell by 15% (Agriculture and Agri-Food Canada 2003c). This trend to larger producers has created a 133% increase in the number of taps per farm and is likely to continue given the costs of equipment and heating (for evaporation) that favor economies of scale.

Further potential for this industry's expansion is demonstrated by data from the Nova Scotia Soils Institute (McIssac 1993) which found that only 10% of the potentially tappable maple stands were being used in 1993. Although expanded production in Nova Scotia may not greatly affect national outputs, it could boost the provincial economy just as it did for Québec. Subsidiary income in rural communities likely will not comprise a significant market trend but can exist in a substantial manner similar to that of blueberries in Ontario and New Brunswick.

Maple products other than syrup, including sugar, taffy and butter, are well known FBF, but consumable products from trees are not only limited to the maple. White birch (*Betula papyrifera*) also produces a unique tasting sap but in much smaller volumes when compared to the sugar maple, as it takes 100 litres of birch sap to make one litre of syrup. Other FBF from birch include juice, (popular in China), beer or wine ("Borealis Birch Beer"), and birch syrup-filled straws, or candy sticks which, according to the Food and Agriculture Organization of the United Nations (Food and Agriculture Organization of the United Nations 2002), are considered healthy products in many countries.

In the early 1990s, Alaska began the first commercial production of birch syrup in North America and today produces approximately 3785 litres per year (University of Alaska 2004). Early success of birch syrup production in Alaska has prompted the Saskatchewan government to consider its potential for that province's northern regions (Saskatchewan Environmental Society 2002). Because birch syrup costs as much as five times that of maple, consumers are typically "high-end". However, interest in birch sap and products appears to be growing around the world, and within several years it is predicted that Alberta will be producing a million litres per year (American Forests 2000), with an estimated value of $31.2 million.

The processing of birch sap can provide an excellent niche activity for communities and individuals seeking subsidiary incomes. Large operations, similar to those for maple sap, are just not profitable when compared with small producers, or "wildcrafters" (Cameron 2001). Collectors in Alaska, for instance, tap birch trees (40,000 gallons in 1998) and bring the sap to a central large processor who can effectively market the syrup. As a lucrative FBF, birch is proving it has the potential to make a significant contribution to Alaska's economy. The potential for opportunities from birch paper also demonstrates another approach for expanding and diversifying these specialized FBF industries.

3.3   Berries

Canada is a world leader in the production of a wide variety of native, forest-based berries, whether grown in their wild state or domesticated. These edible fruits include, but are not limited to, lowbush and highbush blueberry, strawberries (*Fragaria* spp.), cranberries (*Vaccinium* spp.), raspberries, saskatoons, currants (*Ribes* spp.), lignonberries (*Vaccinium vitis-idaea*) and elderberries (*Sambucus* spp.). They can be eaten raw, or processed into a variety of juices, sauces, alcoholic beverages, jams and jellies. Many of the berries are suited to cool northern climates, and it is these more remote locales, on unspoiled lands free of pesticides and herbicides, that drive their labeling as certified organic produce (Agriculture and Agri-Food Canada 2003d).

Domestic and international demand for berries is on the rise with increased public awareness of the many health benefits associated with berry consumption, including high fibre content and antioxidant activity. Research has shown that berries offer good biomedical benefits, and therefore can be promoted vigorously as a health food (Agriculture and Agri-Food Canada 2003e). Canadian acreage of cultivated berries has more than doubled in western Canada during the past decade (Agriculture and Agri-Food Canada 2003e) and future production potential looks optimistic.

Specifically, Canada is the world's largest producer of the wild (lowbush) blueberry (*Vaccinium angustifolium*) producing 43,511 tonnes in 2002 with a FGV of $42M and export value of $106M. Most all lowbush blueberries originate in the glacial soils and northern climates of eastern Canada, with the bulk of commercial harvests occurring in Quebec and Nova Scotia. The US state of Maine also harvested an approximate 28,267 tonnes of wild blueberries that same year (Yarborough 2003).

Approximately 85% of total blueberry production area in Canada is lowbush (as compared to the domesticated highbush), with about 95% of these wild berries sold as a frozen product. It is recognized that most berries freeze well using Individual Quick Freeze (IQF) technology and maintain good product quality for up to two years (Agriculture and Agri-Food Canada 2003d). Alternatives to freezing involve an emerging technology called "infusion", which dries the berry while infusing it with sugar. This creates a very versatile and palatable product that can be readily used by large industrial processors for products like cereals or snack foods (Agriculture and Agri-Food Canada 2003e). Considering the short shelf life of fresh berries, these technologies have proven to be a boon to the industry.

The highbush blueberry (*Vaccinium corymbosum*) represents a fully transitioned FBF where, a century ago, White and Coville domesticated the species so that farmers could commercially grow it. Unlike its wild cousin, the highbush blueberry is grown in milder climates where the wild blueberry is not found. British Columbia boasts 97% of Canadian production (Agriculture and Agri-Food Canada 2003e), and US production occurs mainly in Michigan, New Jersey and Oregon. Production of highbush blueberry in B.C. and other parts of the world is expected to expand, which adds competition to the wild blueberry market, while also singling out their differences for consumers.

At present, the economic contribution of all varieties of berries, both wild and cultivated, to the Canadian economy is $214M farm gate value (FGV) with a total harvest of 149,373 tonnes. With exports included, this figure rises to $278M FGV. Table 5-6 summarizes the current value of most of Canada's significant production including both raw and processed berries.

Table 5-6. Estimated current economic value of berries in Canada.

Commodity	Production ('000)		Economic Value ($'000)		
	Volume (tonnes)	Area (ha)	Farm Gate Value[1]	Export Value[2]	Total Value[3]
Lowbush Blueberries[4]	43,511	40,630	42,425	106,700	106,700
Highbush Blueberries[4]	20,083	3,337	47,290	42,600	47,290
Strawberries[4]	25,712	5,874	52,396	1,000	52,396
Cranberries[4]	43,284	2,942	33,285	30,300	33,285
Raspberries[4]	14,719	3,792	28,155	4,900	28,155
Saskatoon Berries[5]	2.064	1.336	9,100		9,100
Currants and Others[6]	0.415		1,238		1,238[8]
Lignonberries[7]	0.246		490		490
Total	149.373		214,379		278,654

[1]Farm Gate Value (FGV) = total revenue to farmers or primary producers.
[2]Includes Secondary products, processing and value added production.
[3]Average gross annual economic value in millions of Cdn dollars.
[4]Agriculture and Agri-Food Canada (2003d). (2002 statistics).
[5]Agriculture and Agri-Food Canada (2003e). (2001 statistics).
[6] British Columbia Ministry of Agriculture, Food and Fisheries (2004). (2002. statistics). Note: Values extrapolated from previous years' data.
[7]Agriculture and Agri-Food Canada (2003e). (1989 – 1999 statistical mean).
[8]Data for British Columbia.

3.4   Mushrooms and Other Fungi

Historically, wild mushroom production across Canada has not been well-tracked. When formal record-keeping was instigated, Statistics Canada grouped all mushroom production data, both cultivated and wild, under one code, HS07095100, in the World Customs Organization "Harmonized Commodity Description and Coding System". (HS codes are currently used by over 100 countries worldwide for customs tariffs and import/export statistics).

It was not until 2002 that Statistics Canada and Industry Canada expanded their HS codes to distinguish between "agaricus species" (under the original code) and "other mushrooms fresh or chilled" (*e.g.*, pine or matsutake, shiitake (*Lentinus* spp.), chanterelles, boletes (*Boletus* spp.) and morels), coded HS07095900 (see Table 5-7).

Table 5-7. Other Mushrooms (HS07095900) Fresh or Chilled - Exports and Imports – 2003.

Canadian Exports and Imports in $Cdn			B.C. Exports by Destination	
Source	Exports	Imports	Country	$Cdn
British Columbia	7936553	1439581	Japan	3863629
Ontario[2]	1111812	338127	US[1]	1916494
Saskatchewan	36631		France	1373461
Nova Scotia	25157		Netherlands	220838
New Brunswick	15168		Norway	191440
Alberta	12419		Switzerland	129514
Manitoba		39594	Germany	112689
Quebec		4417	Luxembourg	56383
Alberta		20	Others	72105
Canada	9137740	1821739	All Countries	7936553

Source: Industry Canada (2005) HS Codes.
[1]Some mushroom producers in B.C. and the US Pacific Northwest ship their product via the US and vice versa (Tedder et al. 2000).
[2]All Ontario mushrooms (HS07095900) were exported to the US in 2003.

As table 5-7 demonstrates, the majority of Canadian mushroom exports originate from British Columbia. Therefore, in any attempt to define the national economic value and potential for Canadian mushroom harvests, the assumption is that most marketed wild mushrooms are harvested in BC and that virtually the entire harvest is exported to Japan, Europe and the United States (Tedder et al. 2000). This is confirmed by Wills & Lipsey (1999), who examined export earnings of mushroom harvesting companies, seven of which control 90% of the export trade (Table 5-8).

Current potential for wild mushroom exports in Canada can further be extrapolated from the data contained in the above two tables, concluding that the HS system for Canadian wild mushrooms exports in 2003 approached 1139.7 tonnes at $43M. The potential thus appears bright for this industry and BC could eventually export 600 tonnes of matsutake annually to Japan, with other food mushroom exports also equaling this amount – a total value of around $100M (Wills & Lipsey 1999). For all of Canada, exports could add up to $115M, demonstrating good potential for expanded production if infrastructure developments are made. Also noteworthy is that medicinal and nutraceutical mushrooms, extracts and products could have greater economic potential for BC than even the wild food mushroom crop (Wills and Lipsey 1999).

Table 5-8. Estimated Current Output of Mushrooms in the British Columbia Economy[1].

Commodity	Production[2]		$ / kg		Export Value[3] (Average)
	Range[4]	Average	Range[5]	Average	
Matsutake	250 - 392	321.0	38.64 - 90.16	64.40	20.672
Boletes	0 - 100	50.0	23.45 - 35.18	29.32	1.466
Chanterelles	187.5 - 750.0	468.8	13.30 - 19.95	16.62	7.794
Morels (fresh)	20 - 225	122.5	52.76 - 64.65	58.70	7.191
Others[6]	50 = 5 dried	5	53.33 - 66.67[7]	60.00	0.300
Total	507.5 - 1472	989.8			37.423[8]

[1]Extrapolated from Wills and Lipsey (1999).
[2]Production in tonnes (fresh).
[3]Average gross annual economic value in millions of dollars. (Note: 2004 Exchange value of 81.52 yen and 0.75 US per Canadian dollar).
[4]Harvest fluctuations range between good and bad seasons.
[5]Range in export pricing.
[6]Others include lobster, secondary boletes, cauliflower, sweet tooth, hedgehog and others.
[7]Price and production range based on dried lobster mushrooms.
[8]Tedder et al. (2000) estimated exports at 28.7M.

Fresh wild mushroom markets are characterized by high margins due to high labor investments and storage losses. Some species either spoil or get tough quickly, thus requiring an efficient, pre-existing chain of relationships linking harvesters to the final consumer. The easiest route at present seems to involve brokers who can guarantee buyers a constant supply from diverse sources; however, pickers typically receive very low percentages of the final wholesale amount.

Throughout Canada, rural forest communities can take advantage of the potential for ecologically sustainable wild mushroom picking by developing markets that are both reliable and flexible. Development plans could include involvement of external organizations such as the International Fair Trade Association. Supporters of the fair trade movement engage in or buy from enterprises that guarantee as large a portion as possible of retail profits go to the primary producers (International Fair Trade Association 2004).

By far the most economically important forest mushroom (noted for its distinctive fragrance) is the pine mushroom, or matsutake with total world sales in the order of $US 900 million, from which less than 3% of production originates in North America (Northwest Botanicals 2004). Demand for the pine mushroom is concentrated in Japan which, in 2001, imported 63.9% and 13.7% respectively from China and Canada (Japan

External Trade Organization 2004). Due to Japanese cultural traditions, a single pine mushroom may retail for up to US$150, where it will be very thinly sliced and consumed over a week by an entire family. Unfortunately, the Canadian species of *Tricholoma* (*T. magnivelare*) receives a lower price because it doesn't resemble the Japanese species (*T. matsutake*) closely enough. Chapter 4, (Agroforestry, 2.2 FFA - shade systems) provides a convincing description regarding potential for increased economies of scale when pine mushrooms are properly managed.

Prices for many other of the Canadian forest wild mushrooms will almost certainly remain high, especially for those species that form symbiotic associations with trees. These epiphyte fungi tap into roots to gain carbohydrates derived from photosynthesis and in turn provide nutrients to the "host" tree. Ecologists in BC are challenged in determining the carrying capacity of the forest for wild mushrooms, since the bulk of the fungi are underground and the mushroom is merely its fruit. Transitioning these mushrooms out of the forest and developing cultivated varieties most likely will be difficult and impractical in the short run, especially in light of the alternative potential for agroforestry developments (Wills and Lipsey 1999).

3.5 Understorey Plants

A suite of native North American perennial understorey plants inhabit the moist, fertile soils of Eastern Canada's shady hardwood forests. Economically important species which are members of the prodigious lily (*Allium* spp.) family include wild leeks (*Allium tricoccum*), domesticated leeks, onions, garlic, shallots, chives and a host of wildflowers and flowering ornamental bulbs. By far the most valuable of the understorey plants is wild ginseng. Others include wild ginger (*Asarum canadense*) and the fern, fiddleheads (*Matteuccia struthiopteris*).

Historically, wild leeks were an important spring tonic to Aboriginal peoples and European settlers, providing a source of vitamin C and fresh greens absent over the winter. Also known as ramps, wild leeks are most numerous in the rich soils of maple and beech (*Fagus* spp.) hardwood forests in southern Ontario's Carolinian forests and the temperate forests of southern Québec, New Brunswick and Nova Scotia. Because the reproductive portion (bulb) is harvested, wild leeks are very susceptible to over-harvesting, especially as the larger plants that are putting out new offshoots are precisely what pickers are looking for (Gagnon 1999). In Québec, the wild leek was listed as endangered in 1995 and as a result, commercial sale is now prohibited (Ville de Montréal 2004).

Because of this sensitivity to harvest pressure, it is not likely that wild harvesting will be commercially viable as an industry, but small enterprises may continue to supply specialty markets. Propagation, planting and protection of densely populated areas, as the City of Montréal is doing for conservation/rehabilitation purposes, can increase availability of wild leeks. However, as a labor intensive operation, it will not likely be justifiable from a business perspective. Nonetheless, wild leek festivals in the US demonstrate the potential importance of ramps at a local level. In western North Carolina alone, more than 1,453 kg of wild leeks were harvested in 2002 by organizers of eight festivals (US Department of Agriculture 2003).

To a much greater extent than wild leek, excessive harvesting practices have devastated wild ginseng. As far back as the 1890s, over-harvesting was the harbinger of commercial cultivation of ginseng in North America, but it was not until a century later that profits emerged as the driving force expanding this industry. Ginseng, either wildcrafted or grown "wild" in agroforests, has retail values now averaging 10 times the price of field-cultivated roots (Agriculture and Agri-Food Canada 2000a).

The ginseng root is held in high esteem by Asians, particularly the Chinese, who regard it as essential in prescriptions to relieve fatigue and increase vitality. In Western medicine, it is considered a mild stomach tonic and stimulant, useful in loss of appetite and in mild digestive disorders.

Although wild ginseng is still much preferred in Asian markets, since the 1980s, domestication of ginseng to over 2000 ha of production has enabled this Canadian FBF to become a successful commodity in both economic and ecological terms. In 2002, 2300 tonnes worth $75.321 M were exported to Asia, (see Table 5-1); the FGV of ginseng was $54M (Agriculture and Agri-Food Canada 2003f), with Canada the largest grower of North American ginseng, accounting for more than 60% of world production (Agriculture and Agri-Food Canada 2003g).

Similarly, wild ginger (*Asarum* L.) is known for containing important medicinal properties, but it is not sold in large quantities. Canadian wild ginger (*Asarum canadense*) is not related to the cultivated Asian ginger (*Alpinia* spp.) found on grocery store shelves, and its culinary characteristics are different. However, when cooked with sugar, it can be used as a substitute for ginger.

Fiddleheads, also known as ostrich ferns, are picked both by individuals and companies who distribute up to 500 kg per week during the spring season to sellers in urban areas. Although relatively rare and seasonal, fiddleheads are in adequate supply and inexpensive throughout the US. As a result, there will not likely be much potential for future market development within Canada (Mohammed 1999).

The fact that each of these understorey plants is found in association with one another (even if harvest dates are staggered) suggests cost savings for wildcrafters, if they are able to manage these FBF within a single area.

Given their ecological fragility, however, wild understorey plants have very little potential as bulk commodities, especially as leeks, ginger and ginseng already have cultivated cousins that produce large volumes at comparatively low prices. Nevertheless, specialized markets do exist that will pay a higher premium for the unique properties and appeal of wild-harvested FBF.

3.6   Wild Rice

Wild rice (manomenand) is the only cereal native to Canada that will grow from a seed year to year and produce a grain large enough to be consumed by humans. Ojibway peoples traditionally sowed "wild" rice (manomin) in shallow waters along the paths of their canoe routes (Vennum 1988) and it was, for centuries, harvested by the Ojibway, Algonquian and Siouan peoples.

In the late 1950s, growing interest in, and demand for, Canadian northern wild rice (*Zizania palustris*) among non-Aboriginal populations led to increasing industrialization and regulation of rice management (Chapeskie 1990). With the passing of the Ontario Wild Rice Harvesting Act (1960), harvesting rights were assigned through a permit system with little regard to pre-existing and extensive Aboriginal use. With the issuing of these private licenses, many Aboriginals lost access to the resource they had managed from time immemorial. Also, many rice fields that they still had rights to were negatively affected by industrial development, particularly the damming of waterways that flooded these traditionally productive fields.

Today this aquatic plant is generally treated by First Nations and non-Aboriginal peoples alike as a valuable FBF, and harvested both in its natural state and with methods similar to those employed in agroforestry. Production of "Canadian Lake Wild Rice" still occurs mainly in northern waters, without agrochemicals and free of any genetic manipulation. This makes it unique as compared to the cultivated "paddy-grown" rice in the US (Agriculture and Agri-Food Canada 2003h), and also allows for labeling under the Organic Crop Improvement Association's certification (Organic Crop Improvement Association International 2005).

In spite of earlier obstacles, some Aboriginal communities have made use of these developments and persevered in their production and marketing of organic wild rice. For example, a First Nation worker cooperative, Kagiwiosa Manomin Inc. located in Wabigoon (northwestern Ontario), employs about seven people and buys green rice from 100-150 Aboriginal wildcrafters, making a small but important contribution to the local economy. Kagiwiosa Manomin produces between 13,625 to 45,417 kg a year of certified organic wild rice (Organic Crop Improvement Association International 2005) that is processed and packaged on site. Although annual sales of such a business may go unnoticed in the national context, the income and entrepreneurial leadership provided to the Wabigoon region are indeed significant. There are lessons to be learned by any government interested in supporting community enterprise, especially among forest-dwelling Aboriginal peoples, to encourage development of such valuable FBF resources and other bioproducts.

As Table 5-9 demonstrates, in 2003, 1,013 tonnes of wild rice production contributed almost $3.5M to the Canadian economy overall, with 79% of exports destined for the US. A huge ten-fold increase in world production between 1975 and 2001 has shown no sign of letting up, and the future potential for this unique Canadian FBF appears enormous (Catling and Small 2001).

Table 5-9. Canadian Wild Rice Production – 2003.

Province	Production in kg	Economic Value-Cdn$
Saskatchewan	291878	1606959
Quebec	21674	139937
Ontario	454400	788495
Nova Scotia	596	2963
Manitoba	164793	485585
Alberta	79865	468396
Canada	1013206	3492335

Source: Statistics Canada 2004. Canadian International Merchandise Trade, HS code 10089010.

Future challenges facing Canadian wild rice producers, however, are as problematic as those of the past, with potentially grave consequences facing the industry. Currently, there are deep-seated fears that the establishment of field trials for genetically modified rice could lead to "genetic pollution" and irreversibly alter natural strains of wild rice (White Earth Land Recovery Project 2004). This is a critical issue and it is imperative for the industry to protect wild rice as an indigenous resource, and as a certified organic commodity.

3.7   Other Food Products

By far the most promising value-added FBF not previously discussed include the raw ingredients used in the herbal tea market. Herbal teas do not contain any real tea leaf but are comprised of blends of a wide variety of flowers, herbs, spices, fruit, berries and other plants which may or may not have their origins from the forest. Some of these combinations include forest-based wild ginger, strawberry, raspberry (*Rubus* spp.), blueberry, ginseng, mint (*Mentha* spp.), dandelion, labrador tea plant (*Ledum* spp.), stinging nettle (*Urtica* spp.), fireweed (*Epilobium angustifolium*), juniper (*Juniperus* spp.) and rose petals (*Rosa* spp.) (Mitchell M. and Associates 1997).

Without a method to properly track the origins of herbal tea ingredients, the Tea Council of Canada (2001) can only estimate that in 2000, herbal tea sales were worth over $28M. Other estimates (Mohammed 1999; Saskatchewan Environmental Society 2002; Mitchell M. and Associates 1997) place herbal tea sales between 5% to 25% of a $390M Canadian market, or $19.5M-$97.5M. Combined with annual double-digit world-wide growth in tea consumption, this appears to be another promising opportunity for marketing unique FBF products.

Similarly, the contribution of two other significant FBF products, edible tree nuts and essential oils, is challenging to estimate. The total commercial export value for tree nuts is best tracked through Industry Canada's HS codes (HS080290) where, in 2004, Canada exported $2.892M to other countries (Industry Canada 2005). The majority of production occurred in British Columbia at $2.088M, followed by Ontario at $0.502M and Quebec at $0.298M. BC harvests mainly hazelnuts (*Betulaceae* spp.) and, in Ontario, chestnuts (*Castanea* spp.) lead in total value (British Columbia Ministry of Agriculture, Food and Fisheries 2004; Ontario Ministry of Agriculture and Food 2002). Large-scale nut production usually occurs in orchards; however, many local growers do not make reports, adding to the difficulty of estimating volumes (Mohammed 1999).

Essential oils are popular in the cosmetic industry and in food flavourings. Quebec is a major supplier of wildcrafted oils obtained from trees including eastern white cedar (*Thuja occidentalis*) leaf, fir needle, birch (*Betula* spp.) bark, black spruce and eastern hemlock (*Tsuga canadensis*) (Mohammed 1999). The commercial value of essential oils in Canada in 1999 was approximately $1M with the food flavoring contribution as yet unknown and undisclosed by entrepreneurs.

4.   FUNCTIONAL FOODS

There is potential for many FBF to enhance their domestic appeal and export success by emphasizing their reputed "natural health" benefits (*e.g.*, the use of honey and maple syrup over refined sugar). These "functional foods", or nutraceuticals, are food components that provide demonstrated physiological benefits and/or reduce the risk of chronic disease above and beyond their basic nutritional functions (Agriculture and Agri-Food Canada 2003i). Although the contribution of FBF to the these markets is currently not separated from conventional agricultural products, recent estimates gauge the demand in Canada to be in the order of $800M (Table 5-1), with numbers ranging considerably depending on how these FBF are defined. Projections for rapid growth is anticipated (between 8% to 14% per year), with the current Canadian market approaching $1.25 billion (Scott Wolfe Management 2002).

5.   CONCLUSION

Although some of the FBF discussed in this chapter are now produced in domestic agricultural settings, each of these plants originates in and/or is dependent on forests for its productivity and continuity, not simply as individual organisms, but as entire species. Fortunately, domesticated FBF products will continue to share a genetic heritage with their wild cousins, and this shared heritage will be an important source of adaptation to changing growing conditions, especially to combat pressure from disease and pests. As a result of these stresses, germplasm could become an important resource that wild forest food products contribute to domesticated production. Again, appropriate research and management of FBF are essential to preserve and benefit from the abundant diversity of these specialty bioproducts originating in Canadian forests.

Chapter 6

NUTRACEUTICALS FROM THE FOREST

Highlights and Fast Facts

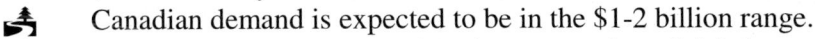	At the turn of the century, the world market for functional foods and nutraceuticals was already $56 billion and is projected to reach $500 billion by 2010.
	Canadian demand is expected to be in the $1-2 billion range.
	Nutraceuticals from Canada's forests are beneficial for providing immune-boosting effects and treating health problems such as high cholesterol and diabetes.
	An aging population in the developed world is demanding natural products to enhance their health and slow the aging process.
	Up-and-coming products include sterols, extracted from wood pulp and used in cholesterol-reducing margarine, and xylitol, extracted from birch and applied as a natural sweetener.
	Progressive marketing of natural health products has been encouraged through recent Canadian legislation.

1. DEFINITION OF NUTRACEUTICALS AND FUNCTIONAL FOODS

The market for nutraceuticals and functional foods worldwide is so significant that these specialized bioproducts warrant their own chapter. As defined in the previous chapter on forest-based foods (FBF), a nutraceutical, or functional food, is any food or food component scientifically proven to provide physiological, medical or health benefits, and/or reduce the risk of chronic disease (including prevention and treatment) above and beyond their

113

basic nutritional functions (Agriculture and Agri-Food Canada 2003i). Forest-based nutraceuticals are products isolated or purified from FBF that are generally sold in medicinal forms, whereas functional foods are consumed as conventional foods, but for the specific purpose of improving or maintaining health (Fitzpatrick 2002).

Nutraceuticals are generally considered part of the vitamin and pharmaceutical market, while functional foods compete for market share with conventional foods (Hobbs 2001). Food and pharmaceutical industry representatives are thus forming joint ventures as they pursue research, development and marketing of nutraceuticals (Culhane 1995).

A roadblock in the development of the nutraceutical industry in Canada, however, has been the lack of specific regulations regarding their official designation. Nutraceutical and functional foods currently are regulated under the Food and Drugs Act. This Act allows for only a limited range of health claims to be made for foods; otherwise they are classified as drugs.

This is problematic, as manufacturers of nutraceuticals and functional foods are reluctant to classify their products as drugs due to the lengthy and costly approval and testing required. Also, from a marketing point of view, companies shy away from positioning nutraceuticals to treat illness, and prefer to promote them as natural products for health and wellness (Health Canada 1998).

As a result, Health Canada is now in the process of developing regulations for the certification of natural health products (NHPs) which collectively consist of vitamins, minerals, herbal products and homeopathic medicines, to mention a few. NHPs are formally classified as drugs under the Food and Drugs Act, from which a full range of health claims derived from these products will be allowed, including structure-function, risk-reduction and therapeutic or treatment claims. The NHP definition consists of two parts: a function component and a substance component. The function components relate to treatment, prevention restoration and maintaining or promoting health through modifying organic functions in humans. (Health Canada 2003). The substance component relies on the fact that the NHP definition is medicinal ingredient-driven.

To further complicate matters, there is little agreement internationally on the definition of nutraceuticals and functional foods. The associated regulations vary from nation to nation, making it difficult for countries with small domestic markets, like Canada, to develop a robust export market (Hobbs 2001). To truly facilitate trade in these products, international guidelines must be established for the labeling of nutraceutical and functional foods, based on clear, irrefutable health benefits and therefore credible to consumers in all countries (Hobbs 2001).

2. HISTORICAL USE OF NUTRACEUTICALS

Plants, fungi, insects and other forest species synthesize or concentrate chemical compounds proven to be effective natural health products, useful in preventing or treating illness, or optimizing body function. These chemicals include sugars, terpenes, aromatics, proteins and alkaloids (Kaufman et al. 1998), and are found in hundreds of living organisms in Canada's forests.

Research into medicinal uses of plants begins with the traditional uses by North American Aboriginal peoples. Native healers have prescreened thousands of forest plants, using many of them for centuries in traditional treatments. As an example, of the more than 7,000 species listed in the Ontario Plant List, 477 are indicated as having medicinal properties as identified by native people (Newmaster et al. 1998). Table 6-1 provides an overview of the Canadian forest plant species that hold potential for deriving nutraceuticals and natural health products.

Table 6-1. Canadian Forest Species With Traditional Medicinal Uses.

Plant Species	Plant Part	Medicinal Use
Acer pensylvanicum (striped maple)	Bark, leaf	Emetic, gonorrhea, kidney infections, skin eruptions
Acer rubrum (red maple)	Bark	Eyewash
Acer spicatum (mountain maple)	Bark, twig	Eyewash, diarrhea, intestinal problems
Achillea millefolium (yarrow)	Flower	Skin inflammation
Agropyron repens (quack grass)	Various	Kidney stones, incontinence
Alnus rugosa (speckled alder)	Bark, leaf	Eye infections, inflammation, ulcers, skin diseases
Amaranthus hybridus (smooth pigweed)	Leaf	Diarrhea, inflammation, ulcers
Amaranthus retroflexus (redroot pigweed)	Various	Canker sores, hemorrhage, inflammation of mouth and throat
Ambrosia artemisiifolia (common ragweed)	Leaf, root	Insect bites, nausea, skin eruptions, stroke
Ambrosia trifida (giant ragweed)	Leaf	Fever, diarrhea, insect bites, mouth sores
Amelanchier spp. (serviceberry)	Fruit, root	Blood remedy for pain and hemorrhage after childbirth
Arctium lappa (burdock)	Root	Hair care and scalp care remedy, cancer treatment as part of Essiac[1]
Asclepias syriaca (common milkweed)	Latex, root	Congestive heart failure, kidney stones, laxative, rheumatism, syphilis, warts

Plant Species	Plant Part	Medicinal Use
Barbarea vulgaris (yellow rocket)	Leaf	Wounds
Betula alleghaniensis (yellow birch)	Bark, twig	Antiseptic, diuretic, local stimulant
Chamaedaphne calyculata (leatherleaf)	Leaf	Fever, inflammation
Chenopodium album (lamb's quarters)	Various	Burns, gout
Cirsium arvense (Canada thistle)	Leaf, root	Diarrhea, diuretic, tuberculosis, skin eruptions
Comptonia peregrina (sweetfern)	Various	Colic, diarrhea, toothache, rheumatism, sores, toxic effects of poisons
Convolvulus arvensis (field bindweed)	Various	Laxative, fever, spider bites, wounds
Cornus alterniflora (alternate-leaved dogwood)	Root	Eye soreness
Cornus rugosa (round-leaved dogwood)	Bark	Tonic
Corylus cornuta (beaked hazel)	Bark, stem	Convulsions, lung hemorrhage, rheumatism, ulcers, teething pain, tumors
Crataegus spp. (hawthorn)	Flower, fruit	Angina pectoris, arteriosclerosis, heart tonic
Diervilla lonicera (bush honeysuckle)	Various	Diuretic, gonorrhea, senility
Epilobium angustifolium (fireweed)	Leaf, root	Burns, inflammation, ulcerous sores
Hypericum perforatum (St. John's wort)	Various	Inflammation, sciatica, sores, urinary disorders, wounds, ulcers, rheumatism, lumbago, upset stomach
Kalmia polifolia (bog laurel)	Leaf	Herpes, sores, ulcers
Lonicera canadensis (Canada fly honeysuckle)	Bark, root, vine	Gonorrhea, lung disorders, urinary disorders
Lythrum salicaria (purple loosestrife)	Entire plant	Antiseptic, diarrhea, sore throat, wounds
Medicago sativa (alfalfa)	Leaf	Inflammation from arthritis, rheumatism, lupus, gout, vitality augmenter
Osmunda cinnamomea (cinnamon fern)	Root	Cough, cold/fever, rheumatism, increased menstruation, intestinal, gastric and liver cancers
Pastinaca sativa (wild parsnip)	Root	Inflammation, pain, sores
Plantago major (common plantain)	Leaf, root, seed mucilage	cough, high cholesterol, sores, wounds, skin injuries, inflammation

Plant Species	Plant Part	Medicinal Use
Populus tremuloides (trembling aspen)	Bark, bud	Arthritis, colds, fever, urinary disorders, venereal disease
Potentilla recta (sulphur cinquefoil)	Root	Diarrhea
Polygala senega (snakeroot)	Root	Pneumonia, bronchitis, purgative
Prunus pensylvanica (pin cherry)	Inner bark, root	Bowel disorders, jaundice
Prunus virginiana (choke cherry)	Root, bark	Bowel disorders, lung disorders, wounds
Pteridium aquilinum (bracken fern)	Root	Burns, headache, promote hair regrowth, sores, insomnia, jaundice
Rheum palmatum Indian rhubarb		Cancer treatment as part of Essiac[1]
Rubus idaeus (red raspberry)	Various	Colds, conjunctivitis, diabetes, diarrhea, alleviates labour pains, sores
Rumex acetosella (sheep sorrel)		Cancer treatment as part of Essiac[1]
Salix bebbiana (beaked willow)	Bark	Diarrhea, fever, inflammation
Salix discolor (pussy willow)	Root, bark	Fever, hemorrhage, inflammation, sore throat
Salix humilis (upland willow)	Root	Colic, fever, hemorrhage, inflammation
Salix petiolaris (slender willow)	Root, bark	Fever, hemorrhage, inflammation, sore throat
Sambucus canadensis (common elder)	Bark, fruit, flower, leaf	Burns, laxative, migraine, wounds
Sambucus pubens (red-berried elder)	Various	Burns, laxative, migraine, wounds
Solidago canadensis (Canada goldenrod)	Flower, root	Burns, fever, snake bites, sore throat
Sorbus spp. (mountain ash)	Various	Antiseptic, colds, colic, pneumonia
Spiraea latifolia (broad-leaved meadowsweet)	Leaf	Diuretic, nausea
Tanacetum parthenium (feverfew)	Leaf	Migraine
Ulmus fulva & rubra (slippery elm)	Inner bark	Cancer treatment as part of Essiac[1]
Vaccinium spp. (blueberry, bilberry)	Leaf, root	Bowel diseases, congestive heart failure, diarrhea, sore throat and mouth, uterine inflammation, cystitis, diabetes
Verbascum thapsus (mullein)	Leaf, flower	Asthma, bronchitis, otitis media, colds, cough

Plant Species	Plant Part	Medicinal Use
Viburnum acerifolium (maple-leaved viburnum)	Inner bark	Cramps, emetic
V. alnifolium (hobblebush)	Leaf	Migraine
V. cassinoides (northern wild raisin)	Bark	Inflammation
V. lentago (nannyberry)	Inner bark	Diuretic
V. trilobum (highbush cranberry)	Inner bark	Colic, cramps, emetic
Xanthoxylum americanum (prickly-ash)	Bark, fruit	Colic, congestive heart failure, expectorant, kidney disorders, lung disorders, rheumatism, sores, topical stimulant, ulcers, venereal diseases

Adapted from: Huang et al. 1999; Mohammed 1999; Latta 1999, Kaegi 1998.
[1]Essiac is a blend of four herbs popularized in the 1920s, used in cancer treatment.

3.   CURRENT TRENDS AND MARKETS

Based on the numerous plant species present (Table 6-1) and their wide-ranging medicinal applications, it is evident that Canadian forests hold abundant potential for producing these valuable bioproducts.

The world market for functional foods and nutraceuticals is large and growing, with recent estimates of market size well over US$50 billion (Government of Canada 2002). Global sales are expected to grow to $500 billion by 2010 (Agriculture and Agri-Food Canada 2003i). The Canadian demand is estimated to be in the $1-2 billion range (Wolfe 2002). In most jurisdictions, the market for nutraceuticals is increasing more rapidly than that of pharmaceutical drugs or food products. Higher prices and greater margins than conventional foods are incentives for companies to enter the market (Wolfe 2002).

The Nutrition Business Journal has identified the primary markets for nutraceuticals and functional foods (and thus Canadian exports) as the US, Europe, Japan and Asia (Fitzpatrick 2002). An estimated 70% of American adults use dietary supplements to improve health and prevent illness, accounting for a US$16 billion market in 2000 (Forbes Medi-Tech Inc. 2004). The Japanese market is approximately US$4 billion.

Japan is a world leader in the area of legislative health claims and food. Their government formally recognizes nutraceuticals as an alternative to drug therapy and has established the regulatory category "Foods for Specific Health Use" (Manitoba Agriculture, Food and Rural Initiatives 2001).

Given the strong market drivers for nutraceuticals and functional foods, the potential demand is enormous. The consumers driving this market are post-World War II "baby boomers" who will not readily accept barriers to their purchasing of natural products to enhance their health and well-being, and slow the aging process (Agriculture and Agri-Food Canada 2002a). Table 6-2 provides an overview of the main areas of health concern among these consumers and outlines the potential of natural health products to address these concerns.

Table 6-2. The Potential of Natural Health Products for Better Health.

Health Concern	Potential Natural Health Products
Joint health	Glucosamine, chondroitin, omega-3 and -6 fatty acids, gamma linolenic acid
Gut remedies	Beneficial bacteria; creative prebiotic and probiotic products; digestive formulae with oligosaccharides and lactobacilli; traditional herbs such as ginger, mint, fennel, papaya, chamomile
Blood fat	Oat bran, psyllium, soy; omega-3, inulin; kiwi extract
Skeletal health	Natural phytoestrogens, inulin, other minerals, soy isoflavones, flaxseed and other herbals
Body fat	Fat burners such as Garcinia cambogia and chromium-containing supplements; fat replacements such as plant sterols
Optimal vision	Beverages and products containing anthocyanins, lutein, zeazanthin, antioxidants to delay aging of eye tissues
Sleep problems, stress, anxiety	Melatonin, valerian, tryptophan, passionflower and chamomile, St. John's Wort
Breast and prostrate cancer	Isoflavones; lycopene; fruits, vegetables, grains and soy; vitamins and herbals
Hormonals	Soy isoflavones, flaxseed, herbals for menopause symptoms; phytoestrogens; natural hormone replacement
Heart health and cholesterol	Phytosterols and derivatives, hawthorn
Mental health and functioning	Ginkgo biloba, St. John's Wort
Skin care and aging	Fireweed, antioxidants, sea buckthorn
Immune system health	Larch, antioxidants, bioflavonoids, ginseng, medicinal mushrooms
Cancer treatment effects	Sea buckthorn, larch, essiac

Adapted from: Sloan (1999).

The heart-benefit functional foods market, valued at US$3.4 billion in 2000, is considered the strongest growth area in the nutraceuticals market, with an expected global market of US$4.7 billion by 2005 (Forbes Medi-Tech Inc. 2002). The market for cholesterol-reducing nutraceuticals was estimated at US$75 million in 2003, with an expected annual growth of 15% (Economist Intelligence Unit 2004).

    This trend may be good news, not only for consumers, but also for the Canadian economy and health care system. Cardiovascular diseases, diabetes and cancer cost the Canadian economy more than $55 billion every year. These major "lifestyle" diseases represent the major and growing component of health care costs, whose total could reach an unsustainable financial burden of $1.4 trillion by 2015 (Conference Board of Canada 2001).

    The Canadian Cancer Society claims that changing diets to include foods and natural products with additional benefits is one of the more simple and effective ways to manage the risks of these diseases (Foragen 2005). Governments, the health care and agri-food sectors, as well as the research community, are therefore enthusiastic about the potential for nutraceuticals and functional foods to improve citizens' health, help growers diversify, and contribute to increased sales of high-value products to niche markets (Agriculture and Agri-Food Canada 2002a; Foragen 2005).

    In 2003, Agriculture and Agri-food Canada sponsored a survey of 576 Canadian companies, providing first-ever information on activities related to functional foods and nutraceuticals. About 17% of these companies had revenues of $10 million or more in 2002, and more than half of all firms exported functional foods and/or nutraceuticals that same year. Export revenues for 11% of these companies amounted to more than $5 million, with another 18% reporting between $1 and $5 million. About 27% of firms reported that the product area generating the highest revenues was "general well-being", while a further 17% cited vascular or heart-health products, and an additional 11%, products related to the immune system.

    Considering this rapid growth in consumer and societal demand, there is abundant room for Canadian forest-derived natural health products to capitalize on this emerging market.

4.   CURRENT FOREST SPECIES USED
AS NUTRACEUTICALS

    Despite this expanding market demand, a significant characteristic of the nutraceutical supply chain is that the botanical raw materials may be in limited supply. To deal with this problem, there has been a shift away from wild-crafted materials towards cultivated crops, which provide a consistent and more reliable supply (Hobbs 2001). However, cultivated crops may not always provide the same potency or capture a specific consumer market as wild-grown ones, as described in Chapter 5.

    Table 6-3 provides an estimate of the current market size of some of the successful nutraceuticals derived from the forest.

Table 6-3. Current Estimated Market Value of Forest-based Nutraceuticals.

Nutraceutical or Food	Estimated World Market $'000	Estimated Canadian Market. $'000
Vaccinium spp.		150,000
Cranberries[1]	666,474	55,410
Ginseng		75,300
Larch arabinogalactan	2,875	
Mushrooms	1,625,000	
Xylitol[2]	30,000	
Plant sterols[3]	70,750	
Elk velvet	100,000	

[1] 1998 figures (Agriculture and Agri-food Canada 2000b).
[2] Hardin (2000).
[3] Based on Raisio Chemicals sales of Benecol products, first quarter (Economist Intelligent Unit 2004).

4.1 *Vaccinium* Species: Bilberry and Blueberry

At least twelve species of bilberry (*Vaccinium myrtillus*) and blueberry grow wild in Canadian forests (Table 6-4). Bilberry is a shrubby perennial related to the blueberry, but the flesh of the fruit is purple rather than white. Bilberry is mentioned in many older texts in Europe and China as an herb valuable for its powerful ability to correct diseases of the digestive system, circulatory system and eyes (Bidleman 2000), and is widely recognized for its ability to promote eye health and vision. During World War II, air force pilots ate bilberry jam to improve their night vision (Alternative Medicine Review 2001).

Table 6-4. Canadian Vaccinium Species and Marketed Products.

Latin Name	Common Name	Marketed Product
V. angustifolium	Low sweet blueberry	Fresh, frozen
V. caespitosum	Dwarf bilberry	
V. corymbosum	Highbush blueberry	Fresh, frozen
V. deliciosum	Cascade blueberry	Fresh, specialty
V. membranaceum	Mountain huckleberry	
V. myrtilloides	Velvet-leaf blueberry	Fresh, frozen
V. myrtillus	European blueberry, bilberry	Extracts, jams, syrups
V. ovalifolium	Tall huckleberry	
V. pallidum	Pale blueberry	
V. parvifolium	Red bilberry	
V. stamineum	Deerberry	
V. uliginosum	Bog blueberry	Tea

Research is currently underway to determine whether any Canadian species have medicinal properties approaching, or possibly exceeding, those of the European bilberry, which is grown commercially. To date, results show that the lowbush blueberries and velvetleaf blueberries (*Vaccinium myrtilloides*) contain 40% to 60% of the anthocyanins found in bilberry (Kalt et al. 1999). Anthocyanosides are powerful antioxidants that prevent free radical damage to collagenous tissue (potentially useful for osteoarthritis, gout and periodontal diseases), and may generally protect the vascular system (Bidleman 2000). These medicinally-important anthocyanins are found in both the peel and the flesh of the bilberry fruit, while only in the peel of blueberries (Kalt et al. 1999).

The largest single market for bilberry is the diabetic population, (6% of North Americans) because it improves circulation, modifies blood sugar levels and helps prevent some of the complications of diabetes (Alternative Medicine Review 2001). Bilberry health products sold in North America are generally in the form of capsule extracts, manufactured in the US and China (the latter using European sources).

One hundred kg of bilberry fruit will produce 1 kg of extract (standardized to 25% anthocyanins). A usual dose of 40-200 mg/day is available from herbal remedy retailers at a cost of about 50 cents. Twice as much blueberry fruit would be required to produce the same amount of extract. Nevertheless, Canadian blueberry growers are looking at ways to benefit from the value of the medicinal and nutritive effects of this plentiful FBF, where approximately 5-30 kg of extract can be produced from one hectare of blueberries, with a value of $25,000-$150,000/ha.

It should be noted that cranberry is another FBF crop with tremendous, but as yet underdeveloped, market potential. Cranberry juice is well known to be effective in preventing urinary tract infections (Avorn et al. 1994). More recent studies indicate that cranberry juice boosts good (HDL) cholesterol by 7.6% and shields the heart with its antioxidant power (Canwest News Service 2004). Cranberry bogs may provide forest communities with another opportunity to diversify their economies.

4.2   Fireweed

Fireweed traditionally has been used to treat skin irritations (Skinner 2002), and recent research has verified ethnobotanical reports of its anti-inflammatory activity (Fytokem 2004). A Canadian company, Fytokem Products Inc., has extracted the bioactive molecule Oenothein-B, patented for its wide-ranging bioactivity, including antiviral and anti-tumor ability (Fytokem 2004). Their line of Canadian Willowherb™ extracts, referred to as dermaceuticals, is marketed to the personal care products industry as ingredients in salves and skin creams for chemical irritation or sunburn.

Fireweed extract has also been shown to effectively kill the bacteria associated with acne. The US market for acne prescriptions is valued at US$2.5 billion (Fytokem 2002). Recent collaboration between Fytokem and a German company with marketing access in Europe and Asia will allow for significant market expansion.

Until recently, most of the supply of fireweed has been harvested from the wild, but there is now limited cultivation. Potential exists for cropping fireweed in commercial intensive forestry, after clearcutting or after prescribed fire.

4.3 Ginseng

Ginseng is a major, highly valuable nutraceutical crop, as discussed in Chapter 5. In 2001, Statistics Canada reported that 380 farms had 2880 ha of ginseng under production (Statistics Canada 2001), and Canadian exports of ginseng were valued at $75.3 million in 2002. Wild-simulated or woods-grown ginseng commands an average price of 10 times that of field-cultivated (Agriculture and Agri-Food Canada 2000a).

Ginseng is noted for reducing stress and fatigue, improving short-term memory, reducing high blood pressure, regulating blood sugar levels, reinforcing the immune system and increasing longevity (Agriculture and Agri-Food Canada 2000a). It traditionally has been favored by the Chinese to improve general health, appetite and memory. It is also commonly used by athletes in the US and Europe to enhance energy. In fact, Japanese medical researchers conducted a study in the mid-1980s showing that ginseng increased total work output on a stationary bike by 23.3%. Russian scientists concurred that ginseng also increased mental performance (Silveron Health Products 2005).

Recently, the ginseng-based product COLD-fX was shown to help prevent infections, reduce the length and severity of symptoms, and cut the number of recurring colds. Significantly, COLD-fX was tested in the first clinical trial of a substance classified as a "natural health product" under the new Canadian regulations (Finlayson 2004).

4.4 Mushrooms

The Asian, European and North American markets for medicinal and nutraceutical mushrooms are growing, with an estimated global market of US$1.3 billion. Most of that demand is from Asia (Wills and Lipsey 1999). Others estimate it to be much higher at US$6 billion (Smit 2003). In Asia, there is a scarcity of some mushroom species due to over-harvesting, deforestation and climatic fluctuations. Because of this, North American mushroom growers' interest lies in the development of species used in traditional Chinese medicine, and biotechnological culturing in a pharmaceutical setting is viewed favorably.

Some producers make powdered ingredients from nutraceutical mushrooms that are used in capsules, health drinks and tablets. The powdered mushroom mycelium sells for an average price of US$100/kg. A number of these products are cited for improving vitality, treating HIV infections, reducing the effects of radiation and chemotherapy, treating cancers, regulating blood sugar, treating Hepatitis B and C, and immune boosting for pets as well as humans (Immunoceuticals Inc. 2004). While there is a disclaimer about the use of these products (due to the fact that they are not approved by medical doctors), numerous scientific studies are provided for the interested consumer to read.

In British Columbia, which contains Canada's most valuable mushroom collection areas, nutraceutical and medicinal mushrooms have greater economic potential than mushrooms collected for food (Wills and Lipsey 1999). Mushrooms native to BC have been shown to contain poly-saccharides, which fortify the immune system. Primarily consumed in Asia, and increasingly in the US, medical research has identified numerous health benefits from the mushrooms' antiviral, anti-tumor, antibacterial and anti-parasitic properties, as well as their general immune-system boosting effects (Table 6-5).

Nutraceuticals can act either as immune stimulants or as adaptogens, which are substances that help the body adapt to environmental and psychological stress by balancing the endocrine system. It appears that certain adaoptogenic mushrooms contain germanium (also found in ginseng), which has been shown to increase the absorption and use of oxygen, and is used by athletes to enhance performance. Other medically important compounds in mushrooms include terpenes, steroids and phytoestrogens (Wills and Lipsey 1999).

The well-known reishi mushrooms (*Ganoderma* spp.) were recorded as a healthy food as early as 300 B.C. (Chilton 1993) and are highly ranked in Oriental traditional medicine as a nourishing tonic. While reishi preparations are used to treat cancer in some countries, only a few clinical trials have been reported in international peer-reviewed journals, and more evidence is likely required before their acceptance by Western oncologists (Smit 2003). The good news is that North American reishi mushrooms are considered of superior quality, because they are organically grown and have a highly accurate standardization of ingredients.

Table 6-5. Medicinal Mushroom Species Found in Canada.

Scientific Name	Common Name	Medicinal Use
Amanita muscaria	Fly Agaric	
Armillaria mellea	Honey Mushroom	
Auricularia auricula	Wood ear	Asian nutraceutical in soups, cholesterol reduction
Boletopsis leucomelaena	Kurotake	
Boletus edulis	King Bolete	
Boletus mirabilis	Velvet Top	
Boletus zelleri	Zeller's Bolete	
Cantharellus spp.	Chanterelles	
Cordyceps spp.	Insect Fungi	Immune enhancing properties, upper respiratory infections, nutraceutical in China
Fomes fomentarius	Tinder Polypore	
Fomitopsis officinalis	Quinine Conk	Cancer prevention & treating digestive tract inflammation
Fomitopsis pinicola	Red Belt	
Ganoderma applanatum	Artist's Conk	Antibiotic properties
Ganoderma oregonense	Varnish Shelf	
Ganoderma tsugae	Hemlock Varnish Shelf	Consumed in Asia as nutraceutical
Gloeophyllum separium	Gilled Polypore	
Grifola frondosa	Maitake	Treatment of AIDS
Hericium abietus	Coral Hydnum	
Inonotus obliquus	Chaga	
Laetiporus sulphureus	Sulfur Shelf	
Lenzites betulina	Gilled Polypore	
Morchella esculentum	Morel	
Phellinus ignarius	Flecked-Flesh Polypore	
Pleurotus ostreatus	Oyster Mushroom	
Polyporus spp (also called *Grifola umbellata*).	Polypores	Cancer treatment & protection from radiation
Schizophyllum commune	Split Gill	Antitumour activity
Trametes versicolor	Turkey Tail	Suppresses spread of tumour cells, health food in Japan
Tremella spp.	Witches Butter	
Tricholoma magnivelare	Pine Mushroom	

Adapted from: Wills & Lipsey 1999; Smit 2003.

Another substance, *C. sinensis*, noted for its remarkable medicinal effects for centuries in China, is derived from *Cordyceps*, a parasitic fungus that grows on insects and spiders, rather than on wood. *Cordyceps* species inhabit Canadian forests and the markets for this nutraceutical in Asia could be met by BC where it is commercially grown under controlled conditions.

4.5   Larch Arabinogalactan

Arabinogalactan is a polysaccharide found in many edible plants, but the major commercial source is the larch (*Larix* spp.) tree. The lower 3-4 meters of the larch trunk have very high concentrations of arabinogalactan, often making it "waste" in timber operations. Larch arabinogalactan was approved by the US Food and Drug Administration in the 1960s as a source of dietary fibre and is receiving increased attention as a useful nutraceutical today. It is thought to increase the level of beneficial intestinal microbes and appears to have therapeutic benefits as an immune-stimulant in preventative medicine. It may also be a useful supplement during cancer treatment (Alternative Medicine Review 2000).

The US company, Larex Inc., is today the primary, if not exclusive, worldwide producer of larch arabinogalactan. It salvages the "waste" wood from timber operations, then extracts and purifies the arabinogalactan through a patented process. It produces 3.6 million kg of arabinogalactan annually at its processing plant in Cohasset, Minnesota. The arabinogalactan is sold to the industrial, health science, health food and supplement manufacturers, with net sales in 2002 of US$2.3 million (Larex Inc. 2003). Larex also markets dietary supplements directly to the consumer in several forms. For example, the Odwalla Company introduced "Glorious Morning" orange juice with cranberries and larch arabinogalactan as a source of fibre and an immune system enhancer (Odwalla 2001). With opportunities like these, there seems to be good potential for a Canadian firm to extract and market arabinogalactan in the western provinces, where an abundance of western larch (*Larix* occidentalis) is found.

4.6   Plant Sterols and Sterolins

Sometimes called phytosterols, plant sterols and sterolins are fats present in plants. They are similar chemically to cholesterol in animals, but are difficult to absorb in the intestine, so their serum concentrations are 800 to 1,000 times lower than cholesterol (Alternative Medicine Review 2001a). Plant sterols may be used in treatments for rheumatoid arthritis, HIV infection, cancer, benign prostatic hypertrophy and diabetes (Alternative Medicine Review 2001a). They also have been shown to safely and effectively lower blood cholesterol levels by inhibiting the absorption of cholesterol in the small intestine.

Various proprietary formulations of phytosterols and phytosterol esters added to the diet have been shown to significantly reduce serum concentrations of LDL-cholesterol in laboratory animals and in humans (Jones et al. 1999; de Graaf et al. 2002). These findings have led to the approval of several products for sale as nutraceuticals and functional foods in Europe, Australia and the US. As noted above, the market for, as well as the potential impact on the health care system from cholesterol-lowering pharmaceuticals, nutraceuticals and functional foods is enormous.

In 1995, Raisio Group of Finland announced that plant sterols extracted from a wood pulp mill could be converted into a fat-soluble stanol ester. They subsequently launched a new margarine, "Benecol", containing 9% plant sterol. The consumption of 25g of this margarine a day can lead to a 10% reduction in cholesterol levels within three to four weeks (Danford 1997). The ease of use and palatability of the margarine as a dietary supplement, and the convenience of supermarket availability, are appealing to many consumers. Benecol sells out in Europe despite a seven-fold price premium. A Vancouver company, Forbes Medi-Tech Inc., manufactures a similar product called "Reducol", with a food-dedicated wood sterol manufacturing plant located in Houston, Texas. It currently has an annual production of 1,000 T of its proprietary phytosterol mixture, with plans to expand to 1,500 T (Forbes Medi-Tech. 2004a).

According to Raisio Group, about 2.3kg of sterols can be extracted per 30 T of pulp (Danford 1997). In 1999, figures for the Canadian pulp and paper industry show total capacity at about 80,000 T/day (Emery 2003). This amount would yield approximately 6 T/day of phytosterols if all were extracted, enough to supply two Forbes Medi-Tech Houston plants. Even so, at a dose of 2 grams/day/person, the Forbes plant can serve only about 1.4 million people.

Of real significance to forest industries in Canada, phytosterols are readily obtained from waste byproducts of forestry operations. Tall oil soap, or pitch, a substance that floats to the top of pulpmaking operations and is a waste product in Canadian pulpmaking, has high concentrations of these chemicals that can be extracted using proven technology. Pine and balsam species provide the tall oil pitch, and northern mills are the best sources (Mohammed 1999). Wood-derived sterols, unlike soy and other agricultural crop-derived sterols, come from non-genetically-modified sources, which is especially important today in the European and Far East markets (Forbes Medi-Tech Inc. 2002).

4.7   Xylitol

Xylitol is a sugar alcohol, termed polyol, which is widely found in plants. Scientists in Finland discovered a process to extract xylitol from corncobs and birch bark after World War II, when there was a sugar shortage. It has been relatively unknown in North America because of the inexpensive supply of cane and beet sugar, but today it is noted for its remarkable health-enhancing properties. There are no known toxic levels of xylitol and it has full approval from the US Food and Drug Administration.

Xylitol is a five-carbon sugar and, unlike the six-carbon sugars predominant in the western diet, is not converted to acids by oral bacteria. It creates an alkaline environment, which is inhospitable to the destructive bacteria responsible for dental caries, therefore helping to prevent periodontal disease. Furthermore, xylitol has 40% fewer calories than sugar and is metabolized very slowly, not causing the abrupt rise and fall in insulin levels that sugar does (Sellman 2002). This makes it the perfect sweetener for diabetics and those with insulin resistance, significant markets in the industrialized world. Currently there are more than 177 million people with diabetes worldwide, and the World Health Organization (WHO) expects that figure to rise to 300 million by 2025 (International Diabetes Federation 2003).

Xylitol is also a useful nutraceutical for people wanting to lose weight. The WHO has extensively documented the need for prevention and control of obesity. The impact of the so-called "obesity epidemic" on non-communicable diseases such as cardiovascular disease, type-2 diabetes and cancer, threatens to overwhelm health systems not only in Canada, but also in many industrialized countries (Foragen 2005). The challenge is to provide citizens with healthier food choices and to motivate them to make the appropriate selections for good health. Such nutraceuticals, therefore, have a role to play in dieting as well as in disease prevention.

Early results indicate that xylitol has other benefits in preventing throat and lung infections, and can inhibit the harmful bacteria that cause ulcers, stomach cancer and childhood ear infections. Xylitol also appears to reverse bone loss, likely due to its ability to promote the absorption of calcium (Sellman 2002).

In the early 1990s, worldwide annual production of xylitol was 5,000 T, with 95% being made by to two firms in Finland (Biswas and Vashishtha 1998). In 2000, the annual Canadian consumption of sugar was 40 kg per capita. Demand for xylitol, therefore, could be up to 40 million kg for the diabetic population in Canada alone. Currently, the major use of xylitol has been in the manufacture of chewing gum, but at a relatively high cost of about $6/kg to manufacturers (Hardin 2000). Its high cost is due to the extraction process, which requires chemical treatment of birch wood fibres under high temperature and pressure, a process that requires an expensive catalyst as well as extensive steps to remove byproducts.

Nonetheless, a number of US companies are interested in large-scale production of xylitol, but likely from corn wastes leftover from ethanol processing. Five hundred T of xylitol can be produced using 17,800 T of bagasse or 6,000 T of corncobs (Biswas and Vashishtha 1998), but similar yields could be obtained from birch chips. Canada also has the potential for xylitol production from residues from forest-based ethanol production, but research would be needed to develop a process and determine the economic feasibility of this extraction. It is likely more feasible to produce xylitol from birch fibre, especially from birch waste harvested from mechanical weed control operations in coniferous plantations.

5.   ELK VELVET

One of the more unusual bioproducts from the forest, elk velvet, is gathered from new live or dead antlers of elk. It has a long history of use in Asia, where it is ingested to increase energy levels, strengthen the immune system, and improve cardiovascular health and brain function (Fitzpatrick 1999). Other health claims include stress reduction, balancing blood pressure, speeding healing and reducing pain from arthritis or injury (Sundance Health Products 2004). It contains chondroitin and gluco-saminoglycan, two compounds known to alleviate arthritis symptoms, as well as a host of minerals, amino acids and other useful compounds. Silveron Health Products sells elk velvet for $15/bottle containing 50 capsules of 280 mg each. This has an equivalent value of about $1,000/kg (Silveron Health Products 2005).

Korea has always dominated the world elk velvet trade, buying 80% to 90% of the exports. Before the Asian financial crisis of the late 1990s, South Korea regularly imported more than 100,000 kg of velvet from Canada, the US, Australia and New Zealand combined (Fitzpatrick 1999). The Canadian market for elk velvet has more recently been affected by the problem with Chronic Wasting Disease among elk herds in western Canada; however, elk ranches can provide yet another opportunity for forest communities to diversify their economy.

6.   CONCLUSION

Interest in nutraceuticals and functional foods is on the rise and Canada is in a unique position to benefit from the anticipated growth. Advantages include a reputation for having a clean forest environment and safe, quality-controlled food products. In addition, Canada has strong research capabilities with collaboration between governments, universities, research institutions and industry. The industry can prosper with the provision of scientific evidence in support of functional ingredients and technologies (Wolfe 2002).

Stakeholders in the nutraceutical industry would benefit from leadership to resolve regulatory issues, reduce the cost of clinical trials and build a communication strategy to enhance awareness of nutraceuticals and their benefits (Culhane 1995). As a progressive example of partnership building to promote the growth of the nascent nutraceutical industry, the Saskatchewan Nutraceutical Network amalgamated with Ag-West Biotech and Bio-Products Saskatchewan to form Ag-West Bio Inc. (Saskatechewan Nutraceutical Network 2004). Saskatchewan aims to boost sales by the nutraceutical industry to $214 million by 2010, based on an annual growth rate of 15% (Saskatchewan Agrivision Corporation Inc. 2002).

Clearly, nutraceuticals and functional foods have great market potential, promise many benefits for citizens, health care systems and economies, and are another source of innovative partnerships and bioproducts from Canada's forests.

Chapter 7

PHARMACEUTICALS FROM THE FOREST

Highlights and Fast Facts

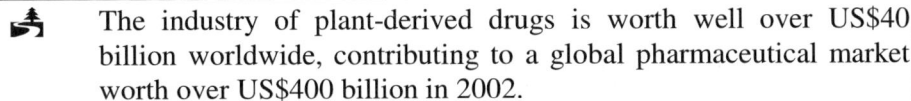

🪑	The industry of plant-derived drugs is worth well over US$40 billion worldwide, contributing to a global pharmaceutical market worth over US$400 billion in 2002.
🪑	Cultivation holds potential for crop diversification and steady supply of plant material necessary for drug development.
🪑	Policies and regulations are needed to ensure sustainable harvesting of plant material and safety of medicinal plant products.

Humans have benefitted from natural sources of drugs for hundreds of thousands of years (Small and Catling 1999). A significant number of Canadian plants used by native people in traditional medicine have found a place in modern medicine, with 466 plant species identified in Ontario alone (Newmaster et al. 1998). There are at least 110,000 known, and possibly as many as a million uncharacterized, natural plant products (R. Williams pers. comm.) serving ecological functions, such as protection against predators and disease. These same products can act within the human body against microorganisms and other causes of illness (van Seters 1995).

The World Health Organization has defined herbal medicines as herbs, herbal materials, herbal preparations and finished herbal products that contain as active ingredients parts of plants, or other plant material, or combinations (World Health Organization 2000). These active ingredients include terpenes, aromatics, proteins and alkaloids (Kaufman et al. 1998). Today, there are over 120 plant-derived drugs in professional use worldwide, three-quarters of which were discovered through scientific investigations of traditional medicines (Marles 1996).

131

Many medicinal plant species are readily cultivated in agricultural or agroforestry settings to provide crop diversification opportunities for producers. Cultivation allows for selection of genotypes that are hardy and produce larger yields, often with higher active ingredient content. Effective marketing of medicinal plant products globally depends on ensuring sustainable supply and harvesting practices, along with regulated standards and quality control (Agriculture and Agri-Food Canada 2000).

To optimize the development of forest-based pharmaceuticals to the economic advantage of First Nations communities, mechanisms should be put in place to allow Aboriginal leaders to share traditional knowledge for societal benefit, while ensuring the protection of their own intellectual property and cultural rights (Bombay 2001).

1.   USE AND POTENTIAL OF FOREST-BASED PHARMACEUTICALS

Pharmaceutical companies, for the most part, use medicinal plants in three ways (Kuipers 1995):

(1) For the isolation of single purified drugs which are made of single chemicals (*e.g.*, digitoxin and vincristine, which are extracted from *Digitalis* and *Catharanthus roseus,* respectively).

(2) In advanced extract forms, where the extract is highly standardized in terms of its content in, or mixture with, active ingredients.

(3) As a starting material for the production of semi-synthetic pharmacologically active substances (*e.g.*, plant saponins can be extracted and altered chemically to produce sapogenins, required to manufacture steroids).

The demand from the pharmaceutical industry for medicinal plants is significant, with roughly 25% of prescription drugs in the US containing plant extracts or active principles prepared from plants (Farnsworth and Soejarto 1985). There are at least 119 chemical substances derived from plants that can be considered as important drugs currently in use in one or more countries (Farnsworth et al. 1985). These drugs are primarily obtained from only about 91 species of plants, most of which could be adapted for cultivation and use in almost every country. Of the 119 chemicals derived from these plants used globally for modern pharmaceuticals, 74% have similar or related uses in traditional medicine (Farnsworth and Soejarto 1985). These findings indicate important potential for new drug discovery pending increased investigation of additional plant species, particularly those with traditional medicinal uses that have not been chemically analyzed to date.

1.1   Pharmaceutical Market Value

Table 7-1 provides examples of plant species and derived pharmaceutical products, illustrating the huge economic potential of plant-derived pharmaceuticals.

Table 7-1. Market Potential of Pharmaceuticals Derived from Canadian Medicinal Plants.

Plant Species	Derived Products	Medicinal Uses	Market Potential
Taxus brevifolia (Pacific yew) & *Taxus canadensis* (Ground hemlock)	Paclitaxel (Taxol®, trade name of Bristol-Myers Squibb Co.)	Treatment for ovarian, breast & other cancers	Worldwide sales of paclitaxel exceeded US$1.6 billion in 2000[1]
Tanacetum parthenium (Feverfew)	Tanacet 125® [2]	Anti-migraine medicine	Over 3 million Canadians suffer from migraine headaches[2]
Epilobium angustifolium (Fireweed)	Canadian Willowherb™ extract (active ingredient, Oenothein-B)[3]	Anti-irritant & potential acne treatment	US acne market (prescription & non-prescription) valued at US$2.5 billion[3]
Podophyllum peltatum (Mayapple)	Teniposide and Etopiside (also known as Vepeside)[4]	Anti-cancer agents	In 1990, the market for drugs from mayapple was worth over $100 million[4]
Rhamnus purshianus (Cascara)	Found in approximately 200 pharmaceutical products in Canada	Laxative	Accounts for 20% of the US laxative market (estimated to be US$400 million annually)[4]
Achillea millefolium (Yarrow)	Found in over 20 pharmaceutical products marketed in Canada	Treatment of gastrointestinal problems & appetite stimulant	Annual world production of the essential oil (800 T) is valued at US$88 million[4]
Wood pulping by-products	Plant sterol based pharmaceutical therapeutics (e.g. Forbes FM-VP4)[5]	Treatment of cardiovascular disease & for lowering cholesterol.	Annual worldwide market for prescription pharmaceuticals for lowering cholesterol was US$15.9 billion in 2000[5]

[1]Phytogen Life Sciences Inc. (2002).
[2]Ashbury Biologicals Inc. (2003).
[3]Fytokem (2002).
[4]Small and Catling (1999).
[5]Forbes Medi-Tech Inc (2003).

Based on the current commercial success and market potential of these products, it is evident that the abundance of medicinal plant species in

Canadian forests holds significant promise for the evolving bioeconomy. Less than a decade ago, the industry of plant-derived drugs was estimated to be worth over US$40 billion worldwide (Marles 1996). More recently, the total world pharmaceutical market was estimated to be worth US $400.6 billion (IMS Health 2003), with the pharmaceutical market in Canada approaching $10 billion annually in sales (Government of Canada 2002).

Despite representing only 1.8% of the world's total pharmaceutical market, Canada accounts for approximately 10% of the new medicines discovered globally (Government of Canada 2002a). New drug development is a lengthy process, and the time lag between discovery and marketing can range from 10 to 20 years (Couzinier and Mamatas 1986; Cragg 1998). However, each new drug is estimated to be worth an average of US$94 million to a private drug company and US$449 million to society as a whole (Mendelsohn and Balick 1995).

Canada has a history as a major world supplier of raw materials, which are exported and then some bought back as finished products. This trend has carried over into medicinal plant crops, which are generally sold in bulk through brokers to American, Asian and European markets (Marles 2001).

Finally, insect-based medicine from Canada's forests is an area of exploration that is almost entirely untapped. Bioprospecting thousands of arthropods could begin with pre-screening species that are known to be medicinally useful to aboriginal peoples. An example of such an insect with apparent health benefits is the June bug (*Phyllophaga* spp.) used by First Nations people to treat anemia and appetite loss after traumatic illness. These insects are prepared by toasting, removing the exoskeleton, and either eating whole or crushed into a powder and mixed with water (Zimmermann 2002).

2.   CANADIAN MEDICINAL PLANT SPECIES

There is a great deal of agricultural, agroforestry and manufacturing potential for numerous medicinal plant species in Canada. Table 7-2 provides an overview of these, as well as information regarding their medicinal applications and market potential. Inventories of these plant species, their suitability for cultivation, as well as research to clarify mechanisms and safe use of their active ingredients, is needed. Value-added, local processing will be required to fully realize their overall commercial potential, and target markets must be clearly identified with stringent quality controls established, to ensure commercial viability (Marles 2001).

Table 7-2. An Overview of Canadian Forest Medicinal Plant Species.

Plant Species	Active Ingredients	Medicinal Uses	Bioproduct Potential
Achillea millefolium (Yarrow)	Lactones (present in a volatile oil), tannins, menthol, camphor, sterols, triterpenes and flavonoids.	Used to treat gastrointestinal problems and as an appetite stimulant.	A relatively undemanding crop that can be grown throughout much of Canada.
Acorus calamus (Sweet flag)	The oil contains hundreds of compounds, particularly phenylpropanes, monoterpenes and sesquiterpenoids.	Used frequently for antibiotic purposes, and useful for relieving digestive disorders and coughs due to antispasm property.	American sweet flag (var. *americanus*) is found in every province in Canada, and has been cultivated in Europe and Asia.
Arctostaphylos uva-ursi (Bearberry)	Leaves contain the phenolic glycoside arbutin (a hydroquinone), monotropein (an iridoid), tannins and trace amounts of aspirin.	Diuretic and anti-inflammatory disinfectant, used to treat bacterial urinary tract infections.	North American medicinal bearberry has been collected from the wild, but is not currently cultivated in Canada.
Arnica spp. (Arnica)	The dried flower heads yield a yellowish-brown powder containing arnican, sesquiterpenoid lactones, volatile oil, resin and tannins.	Primarily used as a counter-irritant (usually as a hydroalcoholic extract) to reduce the inflammation and pain of bruises, sprains and aches.	Most plant material is collected from the wild; however, due to ease of cultivation, arnica is a potential diversification crop.
Astragalus membranaceus (Astragalus)	Roots contain polysaccharides, saponins, flavonoids, phytosterols and essential oils.	Immunostimulant, diuretic antimicrobial and cardiotonic properties; prescribed to cancer patients to increase production of white blood cells; also used to treat chronic diarrhea, high blood pressure and the common cold.	Native to China, but crops are grown in Saskatchewan. Root yields of 7,000 kg/ha are possible. Dried root sells for $8-12/kg for organic product[1].

Plant Species	Active Ingredients	Medicinal Uses	Bioproduct Potential
Caulophyllum spp. (Blue cohosh)	Roots and rhizomes contain glycosides (caulosaponin), and alkaloids (methylcytisine, anagyrine, baptifoline and magniflorine).	Used traditionally to reduce labor pains, regulate menstruation and reduce spasms. Plant extracts have exhibited ovule-inhibitory action in rats, suggesting contraceptive potential.	Blue cohosh has potential for cultivation as a medicinal crop, and the climate and soil in parts of southern Canada are well suited to its growth.
Cimicifuga racemosa (Black cohosh)	The rhizomes and roots contain triterpene glycosides, resin, salicylates, isoferulic acid, sterols and alkaloids.	Used to relieve depression, tinnitus and menstrual cramps. Claimed to be effective as an alternative to hormone replacement therapy for some symptoms of menopause.	The plant is rare in Canada, but is cultivated, and represents an interesting crop diversification opportunity.
Echinacea pallida (Purple coneflower)	A wide array of constituents may contribute to its medicinal properties, with the polysaccharides and alkylamides thought to be most active.	Extracts have cortisone-like antibiotic effects, antiviral properties and a potential for stimulating the immune system.	Wild plants are becoming difficult to find, and cultivation is becoming increasingly necessary to supply the demand.
Epilobium angustifolium (Fireweed)	Myricetin 3-O-ß-D-glucuronide is a novel flavonoid found in the foliage.	Anti-septic and anti-inflammatory uses. Potential as acne medicine[3].	Cultivation for medicinal use could become important if demand increases.
Hamamelis virginiana (Witch hazel)	Tannins and other astringents, as well as flavonoids, may have a curative effect.	The most common usage is in skin lotions to treat varicose veins, sunburn, chapping, insect stings and bites. There is some indication of value for treating aging or wrinkling skin.	Most of the world's supply is obtained from wild plants in the eastern US; however, cultivation appears to have considerable promise.
Hydrastis canadensis (Goldenseal)	Medicinal alkaloids include hydrastine, berberine, berberastine, canadine, and some minor alkaloids.	Used to treat nasal congestion, mouth sores, eye infections, ringworm, hemorrhoids, acne, and as a surface antiseptic. It has a reputation for boosting the immune system.	Much of the current supply is from wild sources; however, the plant is widely cultivated. Holds potential as a diversification crop in southern Canada.

Plant Species	Active Ingredients	Medicinal Uses	Bioproduct Potential
Hypericum perforatum (St. John's Wort)	Naphthodianthrone (mainly hypercin & pseudohypercin), flavonoids, essential oils and tannins.[4]	Used to treat mild depression and anxiety; also used to treat ulcers, bowel inflammation, wounds and bruises; currently under investigation for treatment of HIV conditions.[4]	In Canada, it is an introduced weed. It is grown commercially in Saskatchewan, on a limited basis.[4]
Oenothera biennis (Evening primrose)	Gamma-linolenic acid (GLA), an essential fatty acid for maintenance of cell functions.	Used to treat eczema, migraines, arthritis, diabetes, benign breast disease, high blood pressure and PMS.[5]	In good market years, several hundred hectares may be grown in Canada.
Oplopanax horridus (Devil's club)	Polyynes found in the inner bark and roots, and possibly sesquiterpenes, sesquiterpene alcohol and sequiterpene ketone.	Used to treat arthritis, rheumatism, stomach and digestive problems, tuberculosis, colds and skin disorders. Potentially useful as an anti-diabetic due to its hypoglycemic property.	Found in old growth forests; there are concerns with loss of genetic variation from over-harvesting. Cultivation should be investigated as a possible solution.
Panax quinquefolius (American ginseng)	Ginsenosides (a large variety of triterpene saponins) are found predominantly in the roots.	Widely used as a preventive medicine; used to treat hypotension, hypertension, stress, insomnia, fatigue, depression, arthritis, diabetes, high cholesterol, bronchitis, some cancers, anemia, impotence and premature aging.	Listed as threatened in Canada. Cultivation is necessary to meet consumer demand and Canadian production of American ginseng is increasing.
Pinus banksiana, (Jack pine)	Proanthocyanidin oligomers (pycnogenol) and flavonoids.[6]	Antioxidant property.[6]	Potential market for pine bark extracts needs to be investigated.
Podophyllum peltatum (Mayapple)	Podophyllum resin or podophyllin from the root of the plant.	Contain anti-cancer agents; used internally as a digestive medicine, and also externally in topical medications for genital warts and some skin cancers.	The active ingredient content is variable in the wild, and therefore cultivation of high-yield varieties holds potential.

Plant Species	Active Ingredients	Medicinal Uses	Bioproduct Potential
Polygala senega (Seneca snakeroot)	Active root constituents are triterpenoid saponins (particularly senegin), phenolic acids, polygalitol, methyl salicylate and sterols.	Used as a stimulant expectorant in cough medicines, and to treat asthma and bronchitis.	75% of the world's wild supply originates from Manitoba's Interlake District. Harvesting of the mature root sacrifices the plant, making cultivation a potentially desirable alternative.
Rhamnus purshianus (Cascara)	Mixture of hydroxyanthracene derivatives (particularly anthraquinone glycosides).	Used as a tonic laxative to strengthen the peristaltic muscles of the intestinal wall; also used to treat digestive complaints and hemorrhoids.	The overall retail value of cascara bark is about $100 million, but demand appears to have diminished recently.
Rhodiola rosea (Roseroot)	Variety of phenolic substances, including the glycoside rhodioloside.	Described as an adaptogen, and used to treat stress-related illness.	Not widely used in North America, but cultivated in some Eurasian countries, and holds promise for increased market demand due to its classification as an adaptogen.
Sanguinaria canadensis L. (Bloodroot)	Benzophenanthri-dine alkaloids (particularly sanguinarine), found mainly in the rhizomes.	Used as an expectorant in cough syrup, and has antimicrobial and antitumour properties.	Bloodroot is not currently in demand; however, extracts are the subject of ongoing pharmacological research, and therefore could have economic potential.
Tanacetum parthenium (Feverfew)	Contains sesquiterpene lactones, of which parthenolide is the most important constituent.[2]	Used as a sedative to treat asthma, arthritis, rheumatism, migraine and gynecological problems.[2]	Native to southern Europe but naturalized throughout North America; recently cultivated as a medicinal crop in Saskatchewan.[2]

Plant Species	Active Ingredients	Medicinal Uses	Bioproduct Potential
Taraxacum spp. (Dandelion)	Taraxacin and taraxacerin, along with inulin, gluten, gum and potassium.[7]	Used as diuretic, appetite stimulant, and for treating liver, gall-bladder and digestive problems; possible use for cancer prevention due to antioxidant content.[7]	The annual value of dandelion sold in Canadian markets sometimes exceeds $500,000.
Taxus brevifolia (Pacific yew)	Paclitaxel, found in most parts of the tree, but is especially concentrated in the inner bark.	Used to treat a variety of cancers.	Projected demands exceed the supply of Pacific yew, and cultivation is one solution to meeting market demand.
Taxus canadensis (Ground hemlock)	Paclitaxel, and other taxanes.[8]	Used to treat a variety of cancers.	Due to the high demand for paclitaxel, high-yield cultivars hold great potential as a bioproduct species (see Section 3.1 for more information).
Vaccinium myrtillus (Bilberry)	Anthocyanidin flavonoids (particularly myrtocyan), as well as tannins, myrtillin and chromium.	Believed to be useful for strengthening the cardiovascular system, improving vision, as an anti-diabetic and to treat digestive disorders, mucous membrane inflammation and urinary tract infections; also useful due to antioxidant property.	North American bilberry is not readily accessible (as the plant grows in mountainous regions), and therefore cultivation is desirable.

Source: Small and Catling (1999), except where referenced otherwise.
[1]Porter and Barl (2000).
[2]Porter et al. (2000).
[3]Fytokem (2002).
[4]Saskatchewan Agriculture, Food and Rural Revitalization (2000).
[5]Manitoba Agriculture, Food and Rural Initiatives (2001a).
[6]Marles (1996).
[7]Porter and Brenzil (2003).
[8]Natural Resources Canada (2003b).

2.1   Ground Hemlock and Yew

Ground hemlock is a shrub found throughout Atlantic Canada and Quebec that grows slowly, and often in small isolated pockets, in the understorey of mature and semi-mature forests (Natural Resources Canada 2003c). It is very important pharmaceutically as it contains the active ingredient paclitaxel, used to treat various types of cancer. The extraordinary therapeutic potency of paclitaxel suggests a new way of assigning direct economic value to non-timber forest products.

Cancer is the second most frequent cause of death in industrialized countries, and improved treatments are urgently needed. Ground hemlock-based Taxol® treatment is expensive, with the drug costing between \$10,000 and \$100,000 for each patient, depending on the number of treatment cycles. Not surprisingly, this drug has a commercial value in the order of \$1 billion annually (Small and Catling 1999).

Demand for bulk paciltaxel is forecasted to increase at a 9.7% compound rate through to 2007 (Chatham Biotech Ltd. pers. comm.). On average it takes 30,000 kg of biomass to produce 1 kg of paclitaxel, so a total of 8.4 million kg is needed to supply this demand if all paclitaxel were produced from harvesting wild ground hemlock. Certainly this demand will exceed sustainable harvest from the wild.

To address this issue, the Canadian Forest Service (CFS), in cooperation with the members of the Eastern Canada Ground Hemlock Working Group, has developed sustainable harvesting trials and harvesting guidelines for ground hemlock (Smith and Cameron 2003). The CFS also is promoting domestication to improve the reliability of supply, product quality and provide employment opportunities (Smith and Cameron 2003).

In Northern Ontario, a project has been undertaken to develop ground hemlock as a cultivated, value-added crop, with the aim of selecting plants that have optimum growing conditions and higher chemical concentrations of paciltaxel (Upper Lakes Environmental Research Network 2004). This project could help to meet the growing demand for paclitaxel, as well as create Canadian job opportunities.

Another interesting option deserving consideration is the use of forested electrical transmission corridors for the cultivation of ground hemlock. Plants that provide particularly high yields of paclitaxel could be grown on the land under power lines, also eliminating the need for vegetation control.

In addition to ground hemlock, the genetic tree stock of wild yew (*Taxus* spp.) in Canadian forests is another irreplaceable reservoir of other possible taxanes that may prove to be precursors of equally promising drugs in the future (Wolf and Wortman 1992). Native Americans traditionally applied yew to treat a wide range of ailments including rheumatism, tuberculosis, gonorrhea and wounds. Yew pitch also was mixed with clarified butter and used to treat cancer, an early foreshadowing of its potential for use by modern western societies (Foster 1995).

Several alternatives to harvesting wild Pacific yew to obtain paclitaxel are being developed, including establishing yew plantations, predicted to radically reduce the demand for wild hemlock. There is also research into breeding high-yield cultivars of *Taxus*, production from *Taxomyces* (a genus of fungus isolated from *Taxus*, which also produces taxol), production using tissue culture, full synthesis and synthesis of chemical analogues (Small and Catling 1999). In the meantime, the chief means of augmenting the supply of paclitaxel obtained from the bark of yew is a partial synthesis starting with extracted chemicals from the foliage of species of *Taxus* (Small and Catling 1999). This is a significant development since the foliage can be harvested without sacrificing the trees.

2.2   Phytosterols from Forest Industry By-Products

Phytosterols are lipid-like compounds found in the cells and membranes of oil-producing plants, grains and trees. A large body of scientific research dating back to the 1950s has documented the ability of phytosterols to block the absorption of cholesterol and reduce blood cholesterol levels. Forbes Medi-Tech, a biopharmaceutical company, has carried out research at the University of British Columbia focused on using forest industry by-products as base materials for the production of pharmaceutical products (Forbes Medi-Tech. Inc. 2003). The company has developed a unique process for extracting plant sterols from tall oil soap, a residue from the paper pulping process. They have also created fermentation technology capable of converting plant sterols to pharmaceutical steroid intermediates.

Forbes is now developing plant sterol-based prescription pharmaceuticals that lower blood cholesterol and treat cardiovascular disease. One such product is FM-VP4, an amphipathic analogue of phytostanol, which has demonstrated dramatic cholesterol-lowering and anti-atherosclerotic properties in pre-clinical trials. Clinical trials are currently underway to establish safety and effectiveness parameters. Economic potential is high considering that the annual worldwide market for prescription pharmaceuticals directed at reducing cholesterol reached US$15.9 billion in 2000 and is growing exponentially (Forbes Medi-Tech. Inc. 2003).

2.3   Medicinal Mushrooms

Chinese and Japanese medical practitioners in the US introduced medicinal applications of fungal species to the American public and, presently, numerous firms are actively marketing species for specific medical treatments (Alexander et al. 2002). Dwarfing the markets for wild food mushrooms, the 1997 world market for wild nutraceutical and medicinal mushroom products was US$1.3 billion (US$900 million in Japan plus the rest of Asia, US$250 million for Europe and US$150 million for North America) (Wills and Lipsey 1999).

As described in Chapter 6, British Columbia (BC) is one of the world's most economically valuable, environmentally pristine sources of nutraceutical and medicinal mushrooms (Wills and Lipsey 1999). Mainstream medical research in Japan, China and the Russian Far East indicates that polysaccharides, terpenes, steroids and other ingredients in many BC indigenous mushrooms have antibiotic, antitumour and antiviral properties, reduce lipids in blood, stimulate the immune system, inhibit the synthesis of prostaglandins (hormones which regulate blood vessel size), extend the survival rates of patients with Hodgkins disease, lymphosarcoma and pancreatic cancer, and alleviate some side effects of AIDS (Wills and Lipsey 1999). There are now almost pharmaceutical-grade requirements for nutraceutical and herbal fungal extracts, and product quality, guaranteed potency and standardization of ingredients are all very important (Wills and Lipsey 1999).

3.   MEDICINAL PLANT HARVESTING AND CULTIVATION

Where a species is being evaluated for drug development, considerable quantities of plant biomass are needed for pre-clinical and clinical investigation (Cragg 1998), leading to a risk of over-harvesting and depletion of wild gene pools. Saskatchewan confronts this risk by requiring a permit for all Special Forest Products harvested for commercial purposes, stating: "improper harvesting techniques can cause severe environmental damage or greatly affect the regeneration of the product harvested" (Saskatchewan Agriculture, Food and Rural Revitalization 2000).

In British Columbia there is a provision for protection of forest-based medicinal plants under the Forest Practices Code, but to date no regulations have been made (Turner and Cocksedge 2001). Thus there is really no limit to the quantity of a plant product that may be removed, or restrictions concerning the methods used in harvesting. The main medicinal botanicals gathered from the wild in BC are St. John's Wort (68,040 kg), Oregon grape (*Mahonia aquifolium*) (9,072 kg), cedar oil (2,772 kg) and devil's club (*Oplopanax horridus*) (less than 2,300 kg), and gross revenues paid to harvesters in 1997 were $2-3 million (Wills and Lipsey 1999). Wildcrafters of wild medicinal herbs sell almost exclusively to US bulk suppliers and large manufacturers of herbal medicines.

As the demand for medicinal plants is on the rise, harvest pressure on the resource is increasing. The commercial success of plant-derived drugs, now an industry estimated to be worth over $40 billion world-wide, has led to a great deal of interest in domesticating medicinal plants as alternative crops (Marles 1996). High plant densities can be produced in localized areas, significantly reducing collection and transportation costs (Mohammed 1999). As well, cultivated material is more appropriate for use in the stringent standardized production of drugs, whether for pure products, extracts or crude drugs (Kuipers 1995).

Competition to cultivation, however, is still fueled by the relatively low cost of wild harvested material that is supplied to the market by commercial gatherers who have incurred no costs for growing the plants. Domesticated material requires considerable agricultural expertise and time (sometimes more than 10 years before the crop is ready for harvesting), as well as land and financial resources, before an income can be derived (Kuipers 1995). In cases where domestication has taken place, whether in Europe, Asia or Africa, the crops chosen are those yielding good economic returns or a high level of resource returns, and are either fast-growing species or plants where a sustainable harvest is possible (Cunningham 1995).

Developing native plant species into new crops does present a unique set of agronomic challenges, but for many economic native plants, selection of appropriate varieties and large-scale agricultural production will be the most viable option for a sustainable harvest (Marles 2001). Knowledge of genetic variation and chemical composition is essential for conservation purposes, as well as for selecting the appropriate plants to improve production (He and Sheng 1995). Of course, growing conditions (soil, sunlight, water etc.) must also be considered.

Table 7-3 illustrates price and yield projections, and estimated market size for domesticated crops of some medicinal herbs in Canada. Certified organic product may prove to be more marketable and bring higher prices, and should be considered prior to cultivation.

Table 7-3. Price and Yield Projections and Estimated Market Size for Domesticated Crops of Medicinal Herbs.

Type of Crop	Price and Yield Projections[1]		Estimated Market Size (ha)	
	Price ($/kg)	Yield (kg/ha)	North America	World
Astragalus (root)	17.64	6,725	NA[2]	NA
Black cohosh (root)	13.23	1,120	NA	NA
Dandelion (leaf)	11.02	560	486[3]	6,880[3]
Dandelion (root)	11.02	1,1120		
Echinacea purpurea (root)	15.43	784	4,047[3]	16,188[3]
Echinacea angustifolia (root)	22.05	784		
Feverfew (leaf)	27.56	560	162	1,295
Ginseng (root)	22.05	1,680	7,285	705,822
Goldenseal (root)	110.23	896	6,880	16,997

Source: Alberta Agriculture, Food and Rural Development (2000).
[1]Prices are for certified organic products.
[2]NA = Not available.
[3]Values provided do not specify plant part.

4.   MARKETING, MANAGEMENT AND POLICY ISSUES

Medicinal herbs and forest-based pharmaceuticals are still considered as relatively new products in the North American marketplace and the market for a particular medicinal plant, unless it is the basis for a major biopharmaceutical, is likely to be a small or niche market (Dey 2001). Naturally, market volatility is a concern of the harvesters and growers of medicinal plants, and the stability of market prices directly affects the profitability of production (Schooley 2003).

Critical marketing activities for a bio-medicinal enterprise include researching the demand for medicinal herbs, determining the product accessibility, quality and quantity for the target market, establishing what processing is required, and deciding how best to enter the market (Dey 2001; Kerns et al. 2002). For instance, organic production of medicinal herbs has proven to increase their acceptance by buyers and may yield higher prices. As the understanding of the uses and benefits of medicinal herbs grows, regulatory requirements also can be expected to evolve.

As with most Canadian natural resources, rather than exporting the crude plant material, it would be worthwhile to expand the production of value-added products locally. This would enhance the economic gains from the plant material and possibly create job opportunities in rural and forest communities. As well, there is a need for guidelines and regulations concerning multiple forest resource use and sustainable harvesting practices for medicinal plants in order to protect species from becoming endangered, and to allow harvesters the ability to operate successfully.

The diversity of plant species harvested, and general lack of knowledge about medicinal plants and fungi among many forest managers, complicate the policy making process (Alexander 2001). Furthermore, many species are used in multiple ways, such as edibles, medicinals, for ceremonial and cultural purposes, and as raw materials for crafts and other decorative items (Emery 2001). Communication and coordination among forest resource users can enable timber, medicinal plant and other bioproduct harvesting to be carried out optimally and concurrently, and prevent conflicts among multiple stakeholders.

As an example, many weed species common to forest sites have been used medicinally, and the coordinated harvesting of these species could reduce the need for herbicides and other control methods, thereby increasing the economic potential and environmental health of a given parcel of forest land (Mohammed 1999).

As more plant species which hold potential for drug development are discovered, the risk of over-exploiting plants that are in high demand is a definite concern. It will become increasingly important for the inventory, implementation and enforcement of sustainable harvesting practices and regulations across the country. Costs necessarily will be incurred to implement and monitor any regulations; however, this expense should be offset by the benefits of conservation of biodiversity and wise use of the resources (Turner and Cocksedge 2001). As an example, the harvesting of pacific yew bark for the drug paclitaxel has resulted in the killing of many trees due to a lack of knowledge of best harvesting practices, and no regulations to enforce them.

Land tenure is another issue that should be given careful consideration in order to protect the livelihood of harvesters. Wildcrafters often harvest on crown lands where they do not possess clearly defined legal rights, resulting in uncertainty and vulnerability. The absence of clear tenure or resource management policy leads to a lack of incentives for wildcrafters to conserve available resources for future use (Thadani 2001). Collaboration with First Nations and integration of their traditional knowledge and practices are therefore essential with regard to policy making, marketing and sustainable development of an industry for forest-derived drugs.

5.   CONCLUSION

This chapter describes the many medicinal plant species that exist in Canadian forests, and demonstrates the vast potential of these species to contribute to the national bioeconomy. The pharmaceutical industry is already highly dependent on plants for drug production, and there remains a great deal of unexplored potential for new drug discovery and economic opportunity.

Domestication of medicinal plant species is necessary in some cases to ensure steady access to material where demand exceeds supply from the wild, as well as to protect against loss of species diversity or over-exploitation. Domestication provides diversification opportunities for the farming sector and can spur job creation in northern forest communities that are in need of an economic boost. Value-added processing would further increase the economic returns, employment and social benefits evolving from the medicinal plant industry.

The management and policy issues related to conservation, protection against over-harvesting, intellectual property rights (regarding traditional knowledge), and health and safety of derived pharmaceutical products must be squarely addressed. A strategy to monitor and inventory medicinal plant species and their active ingredients also must be implemented for Canadians to reap the unique benefits of pharmaceuticals from the forest.

Chapter 8

DECORATIVE AND AESTHETIC PRODUCTS

Highlights and Fast Facts

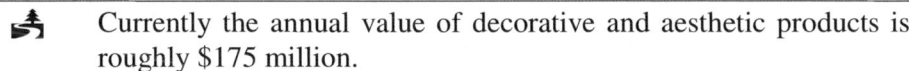	Currently the annual value of decorative and aesthetic products is roughly $175 million.
	The two most significant Canadian decorative bioproducts are Christmas trees ($90 million annual value) and salal ($45 million).
	Production of wreaths and other arts and crafts can be an important source of income for rural families, supplementing annual household income by up to $10,000.
	Traditional native crafts and products made from forest plant species are becoming more highly valued around the world.

1. OVERVIEW

Ornamental forest bioproducts such as Christmas trees, floral and greenery decorations, native arts and crafts, specialty wood products and essential oils do not contribute greatly to Canada's gross domestic product. However, they do have a significant impact on the welfare of certain northern, rural and First Nations communities. Sales of these products can allow families to increase their annual income by $8,000-$10,000 (Duchesne et al. 2001a).

Harvesting activities include the collection of boughs for wreath making, as well as a wide variety of wood, plants and plant parts for craft design. Native communities in particular have a long history of collecting these plants, and some of their traditional functional items, such as ash baskets, are becoming popular because they embody this cultural tradition. In fact, the collection of species such as salal has become so popular that there is now concern about over-harvesting.

Estimating the value of all decorative and aesthetic products emanating from the forest is a difficult task because these figures are not well tracked. Table 8-1 indicates that Christmas trees (including tree plantations cultivated in an agricultural setting), along with greenery collected in British Columbia and balsam fir wreaths produced in New Brunswick, together are worth $175 million. However, many other products such as arts and crafts, specialty wood products and essential oils also make a significant economic contribution. Native crafts, which are largely comprised of items gathered from the forest, are likely worth millions (Mohammed 1999).

Table 8-1. Value of Christmas Trees and Greenery in Canada.

Product	Value ($ millions)
Christmas trees	90
Greenery (including salal)	60[1]
Wreaths	25[2]
Total	175

[1]for B.C. alone.
[2]for N.B. alone.

2.   CHRISTMAS TREES

The Christmas tree industry is an important source of revenue for many Canadian farmers, with 98% of trees grown on land that is not suitable for other uses. Large producers establish wholesale outlets or buyers, whereas those close to urban centres sell directly to the customer or have a "cut-your-own" business, increasingly popular among consumers. Table 8-2 shows the area in production and the number of Christmas tree farms by province.

Table 8-2. Canadian Christmas Tree Production in 2001 by Province.

Province	Area (ha)	# of Farms	% Share
N.S.	9,490	535	25.2
Ont.	8,808	918	23.4
Que.	8,695	395	23.1
B.C.	6,018	526	16.0
N.B.	2,928	252	7.8
Others	2,000	307	5.3
Canada	37,939	2,933	100.0

Source: Agriculture and Agri-Food Canada (2002b).

Despite competition from the "artificial tree" market, close to 6 million Christmas trees are harvested annually in Canada. The industry is worth about

$90 million, with nearly half of this amount generated from exports. In 2001, Canada exported 2.7 million Christmas trees, worth $43.8 million, to about 20 countries around the world, with 98% going to the US (Agriculture and Agri-Food Canada 2002b).

Six species account for 90% of the Canadian and American trade, with scots pine (*Pinus sylvestris*) and Douglas-fir (*Pseudotsuga menziesii*) being the most popular (40% and 30% respectively), followed by noble fir (*Abies procera*), white pine (*Pinus strobus*), balsam fir and white spruce. The main challenge facing the export market is the requirement of "phytosanitary certificates" to ensure pest free material, provided by the Canadian Food Inspection Agency to at least 30% of exported shipments. The total value of exports is outlined in Table 8-3.

Table 8-3. Value of Christmas Tree Exports in 2001 by Province.

Province	Value of Exports ($ millions)	% of Market Share
Que.	25.7	58.7
N.S.	11.0	25.1
N.B.	6.3	14.4
Others	0.8	1.8
Total	43.8	100.0

Source: Agriculture and Agri-Food Canada (2002b).

3. FLORAL AND GREENERY PRODUCTS

Plants collected for the floral industry such as salal, ferns and moss, and conifer boughs for wreath-making constitute a substantial industry with significant opportunities for further development (Mohammed 1999). Plants can be harvested year-round except when winter snow cover impedes access and during the spring growing season, when plant tips are very sensitive. Exports of "foliage, branches, other plant parts, mosses and lichens for bouquets and ornaments", a broad category of products tracked by Statistics Canada, was $74.2 million in 2003 (Industry Canada 2004a). The total number of commercial harvesters is estimated to be in the 12,000-15,000 range (Wills and Lipsey 1999).

Salal comprises 95% of all floral greenery exported from Canada (Ross 1998). It is a low-growing shrub whose branches are often included in flower arrangements. It has shiny evergreen leaves, small white or pink urn-shaped flowers, and edible red or purple berries. Salal grows well in a wide range of soil types and habitats, from open coniferous forests and clear-cuts to the seashore. Its leaves and twigs are browsed by deer and elk, which also use the cover for bedding, and the berries are the favoured food of birds and bears. Many types of small game also use salal for hiding and escape cover.

In British Columbia alone, salal is the second largest non-timber forest product industry, with estimated annual sales at $45 million (Wills and Lipsey 1999). Seventy-five percent of BC salal is exported to Asian markets.

Because it has an extensive, suckering root system that helps bind the soil, salal is also very useful for reclamation purposes in disturbed sites. Further study of the biology of this plant, as well as traditional aboriginal ways of collection and protection, is needed to ensure its sustainability for both ecological and commercial purposes.

Conifer boughs are collected largely for the Christmas wreath industry, but also for cemetery and grave decorations. In New Brunswick, the balsam fir wreath industry began as a cottage industry, but the province is now Canada's largest wreath manufacturer, exporting over 3 million wreaths annually worth over $25 million, and growing at a rate of 15% annually (Falls Brook Centre 2003). In 1998, 9300 tonnes of tips were harvested, which in turn were made into 4.1 million Christmas wreaths worth $20.5 million, creating 4,150 seasonal jobs (Powell 2001). Decorated wreaths also offer an opportunity for greater returns, as the commodity then becomes a craft item. Constraints to foreign exports do exist, such as the need to place fresh wreaths in quarantine.

In contrast with the Christmas greenery market, which is obviously seasonal and mostly within North America, floral greenery markets are year-round and global. Moss, salal and evergreen huckleberry (*Vaccinium ovatum*) are popular with florists in the Netherlands and Germany (von Hagen and Fight 1999). Overseas markets became increasingly important during the 1980s and now the Pacific Rim and Europe have strong potential, especially if an efficient means of product preservation were developed (Schlosser et al. 1991).

Products rise and fall in popularity depending on current tastes in the floral industry. Table 8-4 lists a number of forest plants used commercially in the floral and craft industries. Plants with a deep green colour and long-lasting evergreen properties are the most desirable.

Table 8-4. Forest Plants Used Commercially in the Floral and Craft Industries.

Plant Type	Common Name	Scientific Name
Flowers	Yarrow	*Achillea millefolium*
	Candystick	*Allotropa virgata*
	Pigweed	*Amaranthus* spp.
	Pearly everlasting	*Anaphalis margaritacea*
	St. John's wort	*Hypericum perforatum*
	Blazing star	*Liatris* spp.
	Red-flowering currant	*Ribes sanguineum*
	Starflower	*Trientalis borealis*
	Cattail	*Typha latifolia*

Plant Type	Common Name	Scientific Name
	Bear-grass	*Xerophyllum tenax*
Ferns	Maidenhair fern	*Adiantum pedatum*
	Deer fern	*Blechnum spicant*
	Sword fern	*Polystichum munitum*
Moss and club moss	Broom moss	*Dicranum scoparium*
	Mountain fern moss	*Hylocomium splendens*
	Cypress-leaved feather moss	*Hypnum cupressiforme*
		Kindbergia oregana
		Kindbergia praelonga
	Interrupted club-moss	*Lycopodium annotinum*
	Running clubmoss	*Lycopodium clavatum*
	Ground cedar	*Lycopodium complanatum*
	Ground pine	*Lycopodium dendroideum*
	Common hairy cap moss	*Polytrichum commune*
		Ptilium crista castrensis
		Rhytidiadelphus loreus
	Peat moss	*Sphagnum* spp.
Lichen	Reindeer lichen	*Cladina rangiferina*
Plants used as floral greens	Hairy manzanita	*Arctostaphylos columbiana*
Conifer boughs	Scotch broom	*Cytisus scoparius*
	Leatherwood	*Dirca palustris*
	Horsetail	*Equisetum* spp.
	Salal	*Gaultheria shallon*
	Honeysuckle	*Lonicera dioica*
	Oregon-grape	*Mahonia aquifolium*
	Tall Oregon-grape	*Mahonia nervosa*
	Bog myrtle	*Myrica gale*
	Boxwood	*Pachistima myrsinites*
	Falsebox	*Pachistima myrsinites*
	Evergreen huckleberry	*Vaccinium ovatum*
	Amabilis fir	*Abies amabalis*
	Balsam fir	*Abies balsamea*
	Grand fir	*Abies grandis*
	Alpine fir	*Abies lasiocarpa*
	Yellow cypress	*Chamaecyparis nootkatensis*
	Juniper	*Juniperus* spp.
Conifer cones	Engelmann spruce	*Picea engelmannii*
	White spruce	*Picea glauca*
	Black spruce	*Picea mariana*
	Lodgepole pine	*Pinus contorta*
	Western white pine	*Pinus monticola*
	Ponderosa pine	*Pinus ponderosa*
	Eastern white pine	*Pinus strobus*
	Western red cedar	*Thuja plicata*
	Mountain hemlock	*Tsuga mertensiana*

Plant Type	Common Name	Scientific Name
	Amabilis fir	*Abies amabalis*
	Grand fir	*Abies grandis*
	Alpine fir	*Abies lasiocarpa*
	Yellow cypress	*Chamaecyparis nootkatensis*
	Engelmann spruce	*Picea engelmannii*
	White spruce	*Picea glauca*
	Black spruce	*Picea mariana*
	Jack pine	*Pinus banksiana*
	Lodgepole pine	*Pinus contorta*
	Western white pine	*Pinus monticola*
	Ponderosa pine	*Pinus ponderosa*
	Red pine	*Pinus resinosa*
	Western red cedar	*Thuja plicata*
	Mountain hemlock	*Tsuga mertensiana*
Hardwood twigs	Alder	*Alnus* spp.
	Birch	*Betula* spp.
	Red osier dogwood	*Cornus stolonifera*
	Tamarack	*Larix* spp.
	Oak	*Quercus* spp.
	Pussy willow	*Salix discolor*

Adapted from: Mohammed (1999); de Geus (1995).

In addition to forest plants, there is a market for potpourri worth $500 million to $1 billion (Mitchell M and Associates 1997). Potpourri is an aromatic mixture containing wood shavings, cones, berries, leaves and mosses from the forest. A wide variety of botanical products and fillers are used, such as juniper berries, wintergreen leaves, rosehips, moss and all sizes of cones. Hemlock is considered a premium cone for potpourri because they are light and have an even shape. Essential oils are added to provide fragrance and typically constitute at least one third of the cost of the mix. It has been suggested (Mitchell M. and Associates 1997) that a Boreal Forest potpourri with a unique look and fragrance, containing nuts, acorns, alder cones, dyed wood chips, birch strips and pine needles, could have its own specialty market niche.

Preserved foliage, mosses, lichens and even trees provide another opportunity for year-round sales, and the market is quite diversified. Products are used in the floral industry as well as by landscape, home and business designers and decorators. Foliage can be preserved, dyed or sprayed to meet market demands. There is direct competition with silk products, but a premium is often paid for products made with "real preserved" components. Value-added products, such as wreaths, also can benefit from the use of preserved foliage.

Finally, the tops of birch trees (6 cm diameter, 2 m long) are popular in North America as "look-alike trees" to which artificial leaves and branches are added, for use in offices and hotels (Krantz 2001). Five to seven tops can be harvested from one large birch stem.

4. ARTS AND CRAFTS

Craft products from the forest are proving to be increasingly profitable, supplying a lucrative giftware market (Arborvitae Environmental Services 1997). For instance, from the American giftware retail sales market (valued at US$28 billion), over US$3.6 billion is from seasonal decorative items (Mohammed 1999). The total market value of forest-based arts and crafts is difficult to determine in the US or Canada, since many products are created and sold in rural settings where there are no records of sales. There is no doubt, however, that these products contribute significantly to individual household economies.

A wide variety of woody and herbaceous species, fresh, dried or preserved, are used in the production of craft products, either as a single species or in various combinations. Whole plants, waste wood, leaves, bark, (including bark peels), cones, flowers, berries, roots, branches and twigs can all be used; thus, consideration should be given to the collection of these items in coordination with normal forest operations.

Bark from dead trees can be used to make decorative flowerpots, mosaics, furniture enhancements, birdhouses and many other products (Krantz 2001). Live birch bark is used to make containers of all types as well as novelty items. Twigs from birch, alder, red-osier dogwood and tamarack are collected for use in hardwood wreaths. Decorated wreaths can be sold for up to four times the price of fresh wreaths (Mitchell M. and Associates 1997).

Cones, another byproduct of normal forest processing operations, are popular in the craft market, particularly for Christmas items. After their seeds have been extracted, cones are sold in bulk by seed processing plants (such as the Ontario Tree Seed Plant in Angus, Ontario). In Ontario alone, 300 tonnes of cones are sold from 16 different species of spruce, pine, hemlock, cedar and larch. Sample prices range from $7.50 per 19 kg bag for black spruce to $15.00 for white cedar and larch (Mohammed 1999). Cones that are damaged during the extraction process (Norway spruce (*Picea abies*), white spruce and often, white pine) can still be used in landscaping or as fuel.

The peak period for cone sales is early to mid summer, allowing craft producers time to prepare for the fall retail market. Larger cones are popular in the floral market, small pinecones are valued for wreath making, and all sizes are used in potpourris.

Forest fungi are also collected for unusual crafts, such as artist conk (*Ganoderma applanatum*), a shelf fungus that grows on dying trees. Conks are removed from trees in the winter to prevent smudging or scars, then dried and used by artists for wood burning, painting and carving. A clean 30-40cm conk sells for US$15-20 (Krantz 2001). Other bracket fungi found on hardwoods such as birch and alder can be used to create "amadou", a leather-like substance (Anon. 2002). These fungi are cut into strips and soaked or boiled until they are soft and flat. The resulting suede-like product can be used for novelty items such as ball caps.

Table 8-5 provides a list of commonly-used materials and craft products.

Table 8-5. Summary of Forest Craft Products (British Columbia).

Plant Material Used	Scientific Name	Craft Product
Bark	*Betula papyrifera*	Baskets
	Cornus stolonifera	Weaving
	Picea engelmannii	Rope
	Picea glauca	Birch bark canoes
	Picea mariana	
	Prunus emarginata	
	Thuja plicata	
	Tsuga heterophylla	
Wood	*Abies amabilis*	Furniture
	Acer macrophyllum	Art pieces (carvings)
	Alnus spp.	Tools
	Betula papyrifera	Ceremonial objects
	Chamaecyparis nootkatensis	Hunting bows
	Picea engelmannii	Totem poles
	Picea sitchensis	Arrow shafts
	Pinus contorta	
	Populus tremuloides	
	Salix bebbiana	
	Taxus brevifolia	
	Thuja plicata	War canoes
Leaves	*Arctostaphylos uva-ursi*	Imbrication materials: added as decoration to bark baskets and other types of weaving
	Carex obnupta	
	Chenopodium album	
	Chenopodium capitatum	
	Typha latifolia	
	Xerophyllum tenax	
Whole plants	*Cornus stolonifera*	Natural dyes used by cloth weavers
	Prunus emarginata	

Plant Material Used	Scientific Name	Craft Product
	Scirpus americanus	
	Picea engelmannii	Baskets
	Picea glauca	Rope
	Picea mariana	Weaving

Source: De Geus (1995).

5. NATIVE CRAFTS

First Nations peoples have always made use of offerings from the forest as part of their culture, whether for practical living, commercial (trading) or spiritual purposes. Traditional approaches are relevant to contemporary efforts to harvest decorative forest products, including those practices that maintain the capacity for growth and promote regeneration. However, there is concern about exploitation by wildcrafters who aren't aware of the cultural significance of plant species or how to harvest them to ensure their sustainability (Turner 2001).

Value-added products from aboriginal cultures are more highly valued today than in the past, and the authenticity and cultural tradition embodied in such products are seen as important by tourists and other purchasers (Turner and Cocksedge 2001). A single paper birch tree, valueless as an industrial timber species throughout British Columbia, can be made into $1,000-worth of masks, bowls, spoons and birch-bark baskets by Gitxsan artisans (Turner and Cocksedge 2001).

Traditional crafts from Ontario's native community also are sold in markets in the US and Europe. Species used include the bark and wood of white birch, white elm (*Ulmus americana*), ash (*Fraxinus* spp.), red-osier dogwood, white spruce, alder, poplar and cedar. Sweet grass, spruce roots and sinew are often worked into items. Both inexpensive souvenirs and high-end art suitable for museums and gallery gift shops are made. Examples of popular products include lamp shades and bases, baskets, miniature decorative canoes, model teepees, quill-decorated boxes, tamarack geese, masks, dream-catchers, carvings, jewelry and Christmas tree ornaments (Mohammed 1999).

The market for native crafts has not been quantified but is likely in the order of several million dollars (Mohammed 1999). At present, demand is greater than supply, as it is often difficult to find enough skilled artisans to carry out the painstaking work. Many of these items are sold through agents and wholesalers who serve the local retail market in a particular area. Once an item crosses over from craft to art, it commands a much higher price.

6.   SPECIALTY WOOD PRODUCTS
AND CARVINGS

Specialty wood products from the forest include carvings, toys, jewelry boxes and musical instruments. A typical logging operation removes only the best wood, leaving behind stumps, knots, burls and tree forks that still have the potential to generate income. It is these unusual, and also "fancy", wood grains that are most valuable to craftsmen where scrap and residual wood are often used.

Woodcarvers tend to favour white pine, American basswood (*Tilia americana*), walnut (*Juglans* spp.), black cherry (*Prunus serotina*), butternut (*Juglans cinerea*), white birch, maple, willow, trembling aspen (*Populus tremuloides*), white spruce, eastern white cedar and hawthorn (*Crataegus* spp.), as well as thick bark (8-13 cm). Diagonal slices of eastern red cedar (*Juniperus virginiana*) are used for decoupage art. Burls from spruce, birch and black ash (*Fraxinum americana*) are highly sought after as they can be turned into very decorative pieces, especially bowls, plus they can be used for veneer in the furniture industry. Diamond willow is particularly valuable for furniture, lamps, walking sticks and other products where a diamond pattern occurs after a fungus attacks a branch on any of a variety of species of willow and a callus forms as the wood continues to grow around the wound.

Other examples of carvings include wildlife, working decoys, paddles, nautical and landscape scenes. Techniques include wood burnings, "intarsia" (wood inlays) and chip carvings (Mohammed 1999). Prices vary from a few dollars for a small carving to thousands of dollars for larger and more intricate pieces. Internationally, the US and Europe have proven to be the most lucrative markets. Table 8-6 provides a list of sample specialty wood products.

Table 8-6. Specialty Wood Products from Ontario Species.

Species Name	Common Name	Use
Abies balsamea	Balsam fir	Craft wood
Acer rubrum	Red maple	Clothes hangers & pins, kitchenware, spinning wheels, spools
Acer saccharinum	Silver maple	Products where strength is not required
Acer saccharum	Sugar maple	Utensils, crafts, toys, bowling pins, musical instruments, smokewood, carving
Asimina triloba	Pawpaw	Inner bark woven into cloth
Betula alleghaniensis	Yellow birch	Sled frames, wheel hubs
Betula lenta	Cherry birch	Baskets, wooden ware

Species Name	Common Name	Use
Betula papyrifera	White birch	Canoes, spoons, dishes, arts & crafts, ice cream sticks, toothpicks, spools, toys, clothespins, carving, moose calling horns, snowshoe frames, teepee coverings
Carpinus caroliniana	Blue-beech	Tool handles, small wooden articles
Carya cordiformis	Bitternut hickory	Handles, sporting goods, fuel, smokewood, bark-in chair seats & backs
Carya glabra	Pignut hickory	Tool handles, skis, wagon parts, textile looms
Carya ovata	Shagbark hickory	Smokewood for hams, gun ramrods
Carya tomentosa	Mockernut hickory	Smokewood for hams, charcoal
Castanea dentata	American chestnut	Barrel staves
Celtis occidentalis	Hackberry	Athletic goods
Cornus florida	Flowering dogwood	Weaving shuttles, spools, small pulleys, mallet heads, jeweler's blocks, golf club heads, chisel handles, wedges, knitting needles, sled runners, hay forks, barrel hoops, wheel hubs, machinery bearings, rake teeth
Fagus grandifolia	American beech	Decorative containers, handles, wooden ware, shoes, clothespins
Fraxinus americana	White ash	Canoes, snowshoes, baskets, sporting goods, agricultural & garden implements, playground equipment, oars & keels of small boats, butter tub staves
Fraxinus nigra	Black ash	Baskets, washboards
Fraxinus pennsylvanica	Green ash	Agricultural implements, sporting goods
Fraxinus quadrangulata	Blue ash	Tool handles
Juglans cinerea	Butternut	Carving
Juglnas nigra	Black walnut	Gunstocks, cradles
Juniperus communis	Common juniper	Vine stakes, incense, crafts
Juniperus virginiana	Eastern red cedar	Mothproof chests, carving, pencils
Larix laricina	Tamarack	Roots used to sew birchbark canoes, knees, stringer & keels of boats
Liriodendron tulipifera	Tulip tree	Canoes, well linings
Magnolia acuminata	Cucumber tree	Slats for venetian blinds
Morus rubra	Red mulberry	Agricultural implements, bark in cloaks, bark ropes, boat building
Nyssa sylvatica	Black gum	Tool handles, gunstocks & pistol grips, chopping bowls, agricultural machinery parts
Ostrya virginiana	Ironwood	Tool handles, small wooden articles
Picea glauca	White spruce	Musical instruments (e.g. guitars & violins), carving, paddles

Species Name	Common Name	Use
Picea mariana	Black spruce	Roots as binding material
Picea rubens	Red spruce	Musical instruments (e.g. guitars and violins), piano surrounding boards
Pinus banksiana	Jack pine	Canoe frames; fuelwood
Pinus rigida	Pitch pine	Light torches
Pinus strobus	Eastern white pine	Carving; matches, heddles of looms, craft wood
Platanus occidentalis	Sycamore	Butcher blocks, wheels, chests and trunks, piano and organ cases
Populus deltoides	Eastern cottonwood	Wooden ware, matches, cigar box linings, barrel staves, ironing boards, trunks
Populus grandidentata	Largetooth aspen	Veneer, matches
Populus tremuloides	Trembling aspen	Matches, chopsticks, carving
Prunus serotina	Black cherry	Professional & scientific instruments, handles, toys, carvings, hairbrush backs, musical instruments, weighing apparatus, spirit levels
Quercus alba	White oak	Barrels for liquid, fuelwood
Quercus macrocarpa	Bur oak	Barrels for liquid soap
Quercus rubra	Red oak	Craft wood
Robinia pseudoacacia	Black locust	Rake teeth, took handles, ladder rungs, wheel hubs, policemen's clubs
Salix spp.	Willow	Carvings
Salix discolor	Pussy willow	Decorative twig products
Salix nigra	Black willow	Barrels, toys, wicker baskets
Sassafras albidum	Sassafras	Small boats
Thuja occidentalis	Eastern white cedar	Canoes, chests, carving
Tilia americana	American basswood	Carving, modeling, food boxes, yardsticks, ropes, woven mats, musical instruments, picture puzzle backs, heavy-duty thread, masks
Ulmus americana	White elm	Barrels, boats
Ulmus thomasii	rock elm	Axe handles, wheel hubs & spokes, flour barrels

Source: Mohammed (1999).

6.1   Special Uses of Walnut Husks

Walnut shells are, surprisingly, one of the world's hardest substances. As such, they may be processed into abrasive compounds for use in industry. For instance, pulverized walnut shell is added to rubber formulations in snow tires to improve winter traction. The shells dig into ice and snow like miniature steel spikes, providing excellent gripping power and handling in winter conditions, with minimal tire and road surface wear. Toyo Tires

Canada Inc. produces such tires which exceed standards set by Transport Canada (Toyo Tire Canada 2003).

Walnut-hull blasting is also used for specialty paint removal applications in automotive restoration, because it does not distort the metal in the way that chemical paint stripping does. The process is slower and dustier, but produces significantly less waste, thereby reducing environmental damage and disposal costs.

California is currently the primary source of walnuts. The temperate climate of southern Ontario, however, is suitable for cultivated walnut plantations. Such nut crop production requires the use of special cultivars. These trees are different from the wild black walnut, which produces a nut with a small kernel and a strong flavour, and whose wood is highly valued.

At present, only about 100 ha is under orchard cultivation in Ontario, due to high establishment and management costs (Leuty 1999). Ontario also currently does not have a processing facility for walnut shells, as a steady source of shells from a large orchard industry would be required for such a venture to be economically feasible. The delicious, high quality nut itself is still the most profitable component of production.

7. ESSENTIAL OILS

Essential oils are volatile aromatic compounds found in the leaves, flowers, seeds, bark, roots and fruits of plants. They are used in products such as soap, perfume, deodorant, shampoo, lotions, air freshener, detergent and cleaning agents. Of the world's top 20 essential oils, only eastern red cedar is grown in Canada (Food and Agriculture Organization of the United Nations 2002). However, Canada is a major supplier of certain specialty, wild-harvested essential oils such as cedar leaf, fir needle, birch bark, eastern hemlock, black spruce twig and needle oils (Mohammed 1999). In addition to conifer species, some northern plants are known to produce pleasing odors from their essential oils, such as wintergreen (*Gaultheria procumbens*) (Arborvitae Environmental Services 1997).

Table 8-7 lists some species from Canadian forests containing essential oils, and their uses.

Table 8-7. Canadian Forest Species Containing Essential Oils.

Species Name	Common Name	Use
Abies balsamea	Balsam fir	Detergents, air fresheners, cleaners, disinfectants
Achillea millefolium	Yarrow	Aromatherapy
Acorus calamus	Sweet flag	Perfume
Betula lenta	Cherry birch	Flavouring, medicant

Species Name	Common Name	Use
Gaultheria procumbens	Wintergreen	Flavouring and dentrifice products
Hierochloe odorata	Sweetgrass	Perfume
Juniperus communis	Common juniper	Aromatherapy
Juniperus virginiana	Eastern red cedar	Room sprays, household insect repellants
Ledum groenlandicum	Labrador tea	Perfume, aromatherapy
Mondara fistulosa	Wild bergamot	Tea, aromatherapy
Picea glauca	White spruce	Detergents, disinfectants, soaps
Picea mariana	Black spruce	Perfume & fragrance industry
Pinus sylvestris	Scots pine	Room fresheners, detergents, disinfectants, soaps & vaporizer liquids
Thuja occidentalis	Eastern white cedar	Scent for closets & drawers
Thuja plicata	Western red cedar	Cold remedies, room sprays
Tsuga canadensis	Eastern hemlock	Detergents, disinfectants, soaps

Adapted from: Arborvitae Environmental Services (1997); Ciesla (1998); Manitoba Agriculture, Food & Rural Initiatives (2001); Marles (2001); Mohammed (1999); Small and Catling(1999); Thomas & Schumann(1993).

The production of an essential oil or plant extract is a complex process that varies, depending on the type and end use of the product. Extraction costs tend to be around 60% to 75% of processor costs (Australian Government 2001). Essential oil manufacturers generally distill their oil right at the growing site, thus preserving the quality and the quantity of the harvest.

The primary method of extraction is steam distillation, whereby steam is passed through hundreds of pounds of plant material in a stainless steel vat to diffuse the oil. It takes an enormous amount of plant material to produce a small amount of essential oil (Whole Foods Market 2004). For example, 100 kg of balsam fir foliage yields about 1 kg of oil (Arborvitae Environmental Services 1997).

Canada's primary source of oil is the cedar leaf, produced in British Columbia, Ontario and Quebec, with Quebec being the largest supplier. One of the biggest operations (Cedres Recycles de L'Outaouais) produces about 27 tonnes of cedar leaf oil annually, most of which is exported, and market value has been estimated at US$ 44/kg (Mohammed 1999).

Cedar oil is used extensively in the pet care industry, in flea collars, deodorants, shampoos and hair detangling products (Mohammed 1999). Essential oils help repel ticks and fleas by interfering with the insects' ability to sense the moisture, heat and breath of a pet, and are favoured over the traditional, more toxic, carbamate products.

Demand for essential oils and plant extracts is largely driven by the food flavouring industry, as well as manufacturers of cosmetics, perfumes, fragrances, other scented products and industrial solvents. They also are increasingly used in aromatherapy. Changing consumer preferences in favour of natural over synthetic substances already has had a strong impact on the pharmaceutical and cosmetic industries, which has translated into growing demand for natural sources. Although total sales today are estimated at less than $1 million, it appears that the potential market value of essential oils is likely much higher (Duchesne and Davidson-Hunt 1998).

World trade in the markets for essential oils and their end-use products is growing, as shown in Table 8-8. Each product is discrete, with its own demand drivers, customer groups, cost conditions, uses and price. As a result, demand is typically very oil/plant-specific (Australian Government 2001). Between 1993 and 1998, global demand for essential oils grew at an average rate of 6.1%, and in 1998, exports of essential oils, related perfumes and flavours were valued at US$ 7,435 million, and imports at US$ 6,811 million.

Table 8-8. World Trade in Essential Oils, Perfumes and Flavours.

Year	Exports (US$millions)	Imports (US$millions)
1986	2,149	2,008
1990	4,122	4,206
1994	5,051	4,802
1998	7,435	6,811

Source: Australian Government (2001).

8.   CONCLUSION

To be a successful entrepreneur in the field of decorative and aesthetic forest products, it is important to gain a working knowledge of seasonal demands, prices, product quality, market trends and industry structure (Savage 1995). It is also necessary to understand the unique nature of niche markets, which require more time and energy in marketing. Business owners must familiarize themselves with processing options and requirements to maximize the value of products (such as cold storage to maintain the quality of floral greenery), allowing greater market accessibility (Turner and Cocksedge 2001).

Research on specific products and issues will be required if the industry is to continue expanding (Savage 1995). Cooperation between stakeholders must be secured, as well as an understanding of the nature of the species harvested, to ensure their sustainability. Entrepreneurs will benefit from suitable education programs that include information on species ecology and marketing, as well as from tax incentives, startup funding and research support (Mohammed 2001).

By placing more emphasis on value-added products, which command a higher price, there is less chance of being affected by wide swings in demand and prices for bulk material. Also, more can be earned with fewer raw materials, leading to less chance of over-harvesting the wild plant stocks (Mohammed 2001). Happily, products that are made from native forest plants and those that embody First Nations' traditions cannot be duplicated elsewhere in the world, boosting the potential for expanding markets.

Chapter 9

RECREATION AND OTHER FOREST VALUES

Highlights and Fast Facts

> 🔥 Canadians value forests for many reasons beyond their economic benefits, livelihood and culture — such as recreation, tourism and history, wildlife viewing and consumption, ecosystem preservation, aesthetic inspiration and spiritual connection.
>
> 🔥 Canadians spend billions of dollars each year ($11 billion in 1996) on nature-related activities and experiences, while American visitors to Canada spend over $700 million on wildlife viewing, hunting and recreational fishing.
>
> 🔥 Canada is in a unique position to take advantage of the increased interest in the ecotourism industry, which provides direct revenues of $165 million in British Columbia alone.
>
> 🔥 The value of wild furs harvested is $23 million and the fur industry contributes $800 million to the Canadian economy, providing employment for 68,000 Canadians.

1.   THE VALUE OF NATURE TO CANADIANS

On a per capita basis, Canada literally has the largest endowment of natural resource wealth in the world. In addition to the responsibility of sustainably managing this legacy for economic and social use, managing for other values is becoming increasingly important. Canadians value forests for much more than timber and consumable, tangible bioproducts. Recreation, tourism and preservation are industries which are often included together with bioproducts as values other than timber to be considered when

163

ascribing value to our forests. Together with commodities such as carbon credits and water quality, recreation, tourism and preservation can be considered non-consumptive bioproducts. Canadians want assurance that forests are being protected today and that future generations will have access to this bequest (Canadian Council of Forest Ministers 2000).

Canada's "natural capital" contributes to the welfare of citizens by meeting a variety of requirements, from subsistence to environmental health, to more intangible, psychological needs. However, the value of goods and services that nature provides to people cannot be easily quantified, making it challenging for policy makers to make decisions about competing resource management issues. To help with decision making, a survey was undertaken in 1996 by Statistics Canada on behalf of federal, provincial and territorial agencies to uncover the role that nature plays in the lives of Canadians, and to quantify nature's provision of socio-economic benefits. Information was gathered on participation in nature-related activities, locales and travel, as well as levels of commitment of time and money.

Survey results showed that Canadians spend large amounts of leisure time on activities that depend on natural areas and wildlife, and that participation in forest-related recreation is on the rise. Many of the experiences, such as hiking, camping, water activities, berry-picking, skiing or wildlife viewing, take place within parks, where frequency of visits can be tracked. In 1996-97, there were 14 million visitors to forested national parks, an increase of 14% over the previous five years (Canadian Council of Forest Ministers 2000). A total of 86.5% of the days spent in outdoor activities were spent in forested areas (Canadian Council of Forest Ministers 2000), demonstrating the qualitative importance and quantitative value of forests in providing recreational activities. In Ontario alone, visitors to provincial parks spent $14.5 million on park fees in 1992 and $300 million on their trips to parks (Wildlands League 1997).

The 1996 survey also showed that 85% of the population aged 15 years and over took part in one or more nature-related activity in Canada. Participants used a total of 1.5 billion days of their time enjoying natural settings, spending a total of $11 billion, including $6 billion for trip-related items, $3.1 billion for special equipment and $1.8 billion for other items needed to pursue the activities (Environment Canada 1999). Outdoor activities in natural areas accounted for the largest category of spending ($7.2 billion), followed by $1.9 billion for recreational fishing, $1.3 billion for wildlife viewing, over $800 million for hunting and the remaining $1.2 billion for other nature-related activities.

The information gathered from these kinds of surveys is a useful tool in the management of our natural resources in areas such as policy review, legislation development, land-use planning, marketing strategies and monitoring processes (Environment Canada 2000). For example:

- Work is currently underway to develop socio-economic indicators of sustainability based on the National Round Table on the Environment and the Economy and the World Bank's concept of national wealth, which measures social, natural and human capital. Estimates of recreational values for water, biodiversity and other components of natural wealth are being determined.
- Environmental goods and services will be added to the national income accounts.
- An estimate of the socio-economic benefits of Canada's natural areas can be determined and support for their management can be increased.
- Economic programs can be designed that will support and enhance Canada's biodiversity and ecosystem health.
- The need to adopt a sustainable lifestyle can be encouraged through programs emphasizing environmental learning.
- The economic importance of Canada's tourism industry emphasizes the need to sustain the country's natural wealth.
- Information on the non-product use of forests such as recreation, nature study and wildlife viewing can be useful in forest management policy.
- The education and consultation process will ensure a more informed public when participating in management planning procedures.

The $11.7 billion spent by Canadians and American visitors on nature-related activities led to contributions of $17.3 billion to gross business production and $12.1 billion to Canada's gross domestic product (Environment Canada 2000). The expenditures also created 215,000 jobs with combined income generation of $5.9 billion, and government revenues of $5.4 billion from goods and services taxes (Environment Canada 2000).

It appears that people are willing to pay considerably more for the enjoyment they receive from nature-related activities. For example, a 1998 Reid study concluded that actual expenditures on wildlife viewing in British Columbia represented only half of what people would be willing to pay (Canadian Council of Forest Ministers 2000).

2. ECOTOURISM

Ecotourism is defined as environmentally responsible travel to relatively undisturbed and uncontaminated natural areas, with the objective of respecting, studying, admiring and enjoying the scenery, its wild plants and animals and cultural features (Saskatchewan Environmental Society 2002a). The economic significance of this industry is growing as more people choose to travel to natural areas, coinciding with the public's heightened interest and support for conservation issues. Undisturbed natural places and wilderness areas are the primary ecotourism attractions in Canada, especially if they include opportunities to view native wildlife and plants (Wildlands League 1997). In addition, Canada has a wide variety of ecosystems and good support infrastructure, as well as a solid international reputation as a spacious, pristine country (Saskatchewan Environmental Society 2002).

Ecotourism is the fastest growing sector of a worldwide tourism market, with estimates of annual growth between 10% to 30% (Saskatchewan Environmental Society 2002). The Canadian market share of growth in this area was estimated by Statistics Canada to be 13% in 1993 (Friends of Clayoquot Sound 2004). Increasingly, tourists are pursuing more meaningful and educational experiences such as exploring nature and culture.

A survey of US travelers in the 1990s indicated that 40% were looking for "life-enhancing" travel that focused on active and learning-based vacations. Activities such as nature viewing, photography, hiking and learning about different cultures fit into this category (Saskatchewan Environmental Society 2002). People are becoming more aware and interested in the environmental as well as the social development issues faced by people in the regions they are visiting (Brubacher 1999).

Interestingly, for each Ontarian who hunts, there are three or more people who participate in non-consumptive wildlife viewing, their activities generating more than double the revenue from that of hunters (Wildlands League 1997). In fact, the number of hunters in Ontario declined by 14% over the 1981-91 decade, while non-consumptive wildlife tourists increased by 8% (Wildlands League 1997).

In 1997, the ecotourism industry in British Columbia employed an estimated 13,000 people and had direct revenues of approximately $165 million (Wills & Lipsey 1999). The industry contributed an estimated $792 million to the provincial economy in 1998 (Friends of Clayoquot Sound 2004).

The value of the Canadian park system cannot be under-appreciated. Parks Canada is the country's largest ecotourism operator, with over 16 million visitors annually (Feeny 2004). In 2002, half of all overseas visitors to Canada visited a national or provincial nature park (Feeny 2004). Statistics Canada reported 65 million visits to national parks and conservation areas combined in 1997 (Canadian Heritage 2004).

Ecotourism has the potential to diversify the economy of logging-based communities. While tree harvesting and ecotourism are seemingly opposite in perspective, ecotourism actually fits well into a forested landscape that combines protected areas and sustainable forest use (Wildlands League 1997). Due to the existing infrastructure created for logging communities, additional ecotourism activities can be promoted based on the cultural and educational aspects of compatible forest industries, such as maple sugar operations, arts and craft making, trapping and managed woodlots (Wildlands League 1997). Developing an ecotourism industry in this context helps to retain people and wealth within the community, and avoids single-sector dependence.

2.1 First Nations and Ecotourism

Aboriginal tourism is another major growth industry in Canada. In 1999, it generated about 12,000 jobs and $250 million, with revenues expected to reach the $1 billion mark in 10 years (Friends of Clayoquot Sound 2004). The profile of a tourist interested in First Nations is the same as that of an ecotourist. There is a well-documented demand by tourists from Europe and Japan for exposure to aboriginal cultures (Saskatchewan Environmental Society 2002). A 1995/96 survey found that for 61% of European travelers, visiting native cultural sites was a priority (Friends of Clayoquot Sound 2004).

Eighty percent of aboriginal communities lie within, or near, productive forest zones, and forests are valued for their economic, spiritual, cultural and traditional uses (Canadian Council of Forest Ministers 2000). The demand for aboriginal and nature-based tours is growing, as high-income, urban visitors want to experience traditional and spiritual values such as those related to wild food harvesting and unique forest experiences (Tedder et al. 2000). In their 1999 economic strategy report, Wills and Lipsey identified ecotourism as holding the greatest promise of all the non-timber forest products and services to bring significant revenue into First Nations communities.

The advantage of ecotourism activities based on knowledge and use of wild plants is that no harvesting needs to take place, avoiding controversy over potential misuse or overharvesting. In Nova Scotia, for example, guided walks related to the traditional use of plants by the Micmac are offered (Brubacher 1999). Interest appears to be growing in activities related to gathering and identifying fungi, as well as in the use of wild-gathered materials in craft making (Tedder et al. 2000).

As the popularity of ecotourism grows, care must be taken to ensure that those aspects of the First Nations culture that are sacred can be protected in perpetuity (Friends of Clayoquot Sound 2004).

3.   FOREST WILDLIFE

Wildlife populations are important both in an economic sense as well as for their role in maintaining healthy ecosystems. Biological processes in which wildlife plays a key role include pollination, germination, seed dispersal, regeneration, nutrient cycling, predation, habitat maintenance, waste breakdown and pest control (Canadian Wildlife Service 2005). Wildlife provides direct benefits to many First Nations communities in the form of food, clothing and income. For some native populations, over half of their total income comes directly from hunting, fishing and trapping (Canadian Wildlife Service 2005).

3.1   The Fur Industry

Wild fur trapping has a long history in Canada, much of it linked to the Hudson's Bay Company. Historically, beavers were the mainstay of the fur trade, but by 1900 their population was much reduced. However, populations have since returned due to successful management programs. Harvest levels tend to vary with price fluctuations and trappers will switch to other species if prices fall. Most fur-bearers have had their population positively affected by trapping. Marten and lynx however, can be over-trapped and therefore must be monitored closely. The key to the future status of many of these species is to maintain their habitat requirements.

Currently, the mainstay of the Canadian wild fur harvest is the muskrat (35%), followed by beaver (22%) and marten (17%). Other species include fox, coyote, squirrel and raccoon. A summary of the most recent figures for fur prices and number of pelts harvested is shown in table 9-1.

The fur industry as a whole contributes $800 million to the Canadian economy (Fur Council of Canada 2004) and provides employment for 68,000 Canadians (as shown in table 9-2), including those in rural and aboriginal communities. Total fur exports, valued at $335 million in 2003, have increased over the last five years (see table 9-3), with 80% of exports going to the United States and the rest to Japan and the European Union. Of the more than 2 million fur pelts produced in Canada each year, just under half of these are wild furs with the rest raised on fur farms (see table 9-4). Currently the value of pelts harvested from the wild, in addition to ranch-raised pelts, is $73.5 million (see table 9-4).

Table 9-1. 2002-2003 Wildlife Harvest Statistics.

Species	Number of Pelts	Total Value ($)
Mink	28,090	479,522
Badger	433	15,937
Marten	107,278	5,156,153
Beaver	164,498	3,718,902
Bear	3,818	493,949
Lynx	10,141	1,376,950
Wolf	2669	335,679
Coyote	45,805	2,392,854
Fisher	18,431	672,086
Fox	41,323	1,958,099
Muskrat	167,318	638,946
Otter	16,122	2,219,765
Wolverine	468	116,063
Raccoon	59,189	924,064
Ermine (weasel)	39,363	161,987
Skunk	526	3,989
Squirrel	92,071	154.017
Wildcat or Bobcat	2,048	424,882
Total	779,591[1]	21,243,844[2]

Source: Statistics Canada (2004b).
[1]Total does not include furs from Saskatchewan, estimated to be 85,530.
[2]Total does not include value of Saskatchewan furs, estimated to be worth $1,907,720.

Table 9-2. Number of People Employed by the Fur Industry.

Employment Sector	Number of People Employed
Trappers	60,000
Fur farming	2,000
Manufacturing & processing	2,500
Retailing	2,500
Related services	1,000

Source: Fur Council of Canada (2004).

Table 9-3. Value of Canadian Fur Exports 1999-2003 ($ millions).

	1999	2000	2001	2002	2003
Raw furs	132.9	175.9	194.1	197.8	211.3
Dressed furs[1]	21.5	20.3	22.1	25.8	31.8
Fur garments	95.7	119.2	118.4	103.4	91.9
Total	250.1	315.4	334.6	327.0	335.0

Source: Fur Council of Canada (2004).
[1] dressed furs are tanned.

Table 9-4. Fur Production in Canada 1998-2002.

Year	Quantity of Pelts (000's)			Value ($ 000's)		
	Wildlife	Ranch-raised	Total	Wildlife	Ranch-raised	Total
1998	1,059	1,003	2,062	16,740	30,500	47,240
1999	1,078	1,064	2,142	18,902	46,210	65,112
2000	878	1,090	1,968	20,638	46,204	66,842
2001	1,059	1,147	2,206	24,234	49,971	74,205
2002	895	1,140	2,035	23,611	49,913	73,524

Source: Statistics Canada (2004c).

3.2 Hunting and Wildlife Viewing

Canada's abundant fauna attracts many foreign visitors. In fact, twice as many Americans visit Canada for wildlife viewing or recreational fishing (1.1 million visitors), spend nearly twice as many days in Canada (6 million days) and three times more money (Environment Canada 2000) compared to the number of Canadians who visit the US for the same reasons. The 1996 nature survey revealed that 4.1% of Canadians hunt wildlife in Canada (Environment Canada 1999), while 18.6% participate in wildlife viewing. A survey by the US Fish and Wildlife service estimated that American visitors to Canada spent over $700 million on wildlife viewing and recreational fishing.

Average revenue collected from hunting licenses for moose, deer, bear and small game amounted to more than $32 million annually for the period 1993-1998 for the provinces of Quebec, Newfoundland and Labrador, Nova Scotia, New Brunswick and Ontario (Canadian Council of Forest Ministers 2000), and is likely double that for all of Canada. A study by the Canadian Wildlife Federation showed that hunters pay almost $70 million annually on hunting licenses and fees (Mauser 2005). Statistics on annual harvest and revenue collected for half of Canada's provinces are outlined in table 9-5.

Table 9-5. Average Annual Harvest and Value of Game Species for Some Canadian Provinces[1] 1993-1998.

Species	Number of Hunting Licenses Sold	Number of Wild Game Harvested	Revenue Collected ($)
Moose	279,935	42,283	12,161,977
Deer	445,287	114,183	12,662,004
Bear	43,149	10,823	2,713,041
Small game	381,615	435,480	5,083,760
Total	1,149,986	602,769	32,620,782

Source: Canadian Council of Forest Ministers (2000).
[1]Includes Quebec, Newfoundland and Labrador, Nova Scotia, New Brunswick and Ontario.

3.3 Value of Wild Herds for Meat

Wild game farming is a relatively new industry in Canada, with the number and size of herds rapidly increasing. Wild game herds were valued at $350 million in 2001, and included four major species: bison, elk, reindeer and whitetail deer. The value of each of these is outlined in table 9-6.

Table 9-6. Value and Size of Canadian Wild Meat Herds, 2001.

Species	Herd Size	Average Value ($ per head)	Total Value ($ 000's)
Bison	144,000	1,105	159,097
Elk	84,887	1,598	135,627
Reindeer	10,345	1,500	15,518
Whitetail deer	24,310	1,658	40,295
Total	263,542		350,537

Source: Ray Nixdorf, Provincial Livestock Production Specialist, Saskatchewan Agriculture, Food and Rural Revitalization (personal communication).

4. CONCLUSION

Canadian forests provide numerous benefits in terms of wildlife habitat and potential for recreation and tourism. Surveys have yielded convincing evidence that there are significant socio-economic advantages to managing timbered lands for other uses and values. The importance of our natural assets to the tourism industry are especially hopeful. Ecotourism appears to be a venture that can provide long-term economic, social and environmental sustainability.

Also, with the increased interest in learning about traditional uses of non-timber forest products comes the potential to develop non-consumptive endeavors that benefit forest-based communities. To reap the potential from these other forest values, healthy forest ecosystems need to be maintained and protected. Canada's parks and natural areas must be recognized for their long-term importance, at the local, national and international levels.

Chapter 10

PROTECTING FOREST BIOMASS FROM PESTS: CURRENT CHALLENGES

Highlights and Fast Facts

- Forest area lost to insect destruction is far more extensive and economically devastating each year than area lost to wild fires.
- Invasive, foreign species (insects, plants and animals not native to Canada) are threatening forest health and biodiversity.
- Up to $1 billion is spent on pest control products in Canada each year.
- There is increasing consumer resistance to the perceived and/or real negative health and environmental costs of using chemical pesticides.
- Biological pesticides are generally seen as less toxic and safer to people and the environment, more pest-specific, and more biodegradable than conventional (synthetic) pesticides.
- Many of the defenses that plants produce against pests are the same natural products that can have health benefits to humans.
- Propagation of trees with "preferred" traits such as disease and insect tolerance, drought tolerance, cold-hardiness, reduced lignin and increased biomass, can provide needed feedstocks for bioenergy production, platform biochemicals, plastics etc., while at the same time alleviate pressures on old-growth and recreational forests.

1.   MANAGEMENT OF FOREST INSECT PESTS – SCOPE OF THE ISSUE

Insects make up the most diversified group of living organisms on the planet. Recent estimates by the Canadian Forest Service have placed the number of insects on Earth at between 5 and 30 million, of which only one million have been discovered. Although insects often wreak considerable economic damage to Canada's forests (and therefore are often considered as "pests"), they are a crucial part of woodland fauna and they reflect the complexity of existing habitats and of biodiversity as a whole. Insects participate in most of the ecological processes that sustain forest ecosystems. This is why researchers have a keen interest in the ecology and physiology of each individual insect pest. Much attention is also given to curtailing the considerable damage (and resulting financial toll) caused by insects to Canadian forests and the forest industry.

As emphasized often in this book, our nation's forests are worth protecting for a variety of critical reasons. A priority for Canada is the central role our forest products industry plays in our national economy, and will play even more so in an emerging bio-based economy. Canada already exports more forest products than any other nation in the world. In 2002, according to figures compiled by the Canadian government and the Forest Products Association of Canada, the Canadian forest industry exported \$43.2 billion worth of products and is the nation's largest industrial employer. The sector is also the most geographically dispersed industrial employer in the country, supporting more than 300 rural communities.

Forest pests and diseases have a significant impact on Canada's forests, causing losses of approximately 45 million cubic meters of wood per year by reducing tree growth, killing trees or causing wood degradation. In particular, insect pests are demolishing trees across millions of hectares of Canadian forest. Rising temperatures and higher concentrations of green-house gases have led to a reproductive increase in mountain pine beetles, spruce budworms, jack pine budworms, tent caterpillars and other pests in climate-changed Canadian forests, with insects taking advantage of milder winters, hot dry summers and ozone/carbon dioxide-induced changes to pest resistance in trees (BIOCAP Canada Foundation 2004).

With the globalization of markets and trade, there also has been a sharp increase in the number of non-native, or exotic insect species introduced into North America in recent years. Several species that were accidentally introduced in the past (*e.g.*, through transport on ships) have caused considerable damage to Canadian forests. The Canadian Forest Service and Canadian Food Inspection Agency, responsible for detecting exotic insects, cite the destruction in British Columbia and Eastern Canada by the Asian gypsy moth (*Lymantria dispar*) to illustrate this situation only too well.

Introduced foreign species (including plants, insects and animals) can quickly modify the biodiversity of our forests and upset the fragile equilibrium of ecosystems. Over the past few years, several exotic insect species have been caught in the Greater Vancouver Area, and some of them, such as bark beetles, have become quite abundant, thereby altering the balance of native ecological communities. The Asian long-horned beetle and similar species have been detected in such diverse areas as in packing wood in the Port of Vancouver, in Halifax's Point Pleasant Park and in Toronto's urban forests (Natural Resources Canada 2004a).

In 2003, in British Columbia alone, the mountain pine beetle infested 4.2 million hectares of interior woodland, more than double the area affected just one year earlier. (In contrast, the infamous 2003 BC wildfires burned an estimated 265,373 hectares.) By 2006, the provincial government expects the pine beetle epidemic to infest about 480 cubic meters of BC timber, more than six times the current allowable timber harvest for the province's entire forest industry (BIOCAP Canada Foundation 2004).

The Port of Montreal is another point of entry for exotic species because of the huge tonnage of merchandise that is handled there. In the last decade, Eastern Canada saw many large-scale hemlock looper (*Lambdina fiscellaria*) infestations in regions that had never been seriously affected before. The Canadian Forest Service considers this insect to be the second most significant forest pest after the spruce budworm (*Choristoneura fumiferana*). Although looper outbreaks are more sporadic and geographically limited than those of the widespread spruce budworm, they are more immediately devastating. In contrast with the budworm, hemlock looper epidemics develop very rapidly, since the larvae attack older foliage (both coniferous and deciduous) as well as the current year's growth, and the host trees die quite quickly.

In the province of Quebec, the hemlock looper has been outdoing all other insects in terms of damage to forests with more than 1 million hectares destroyed since the early 1990s. In 1999 alone, 472,000 hectares of forest representing over 23 million cubic meters of wood were infested, which is nearly 30 times the area of forest defoliated by the spruce budworm during the same year. Canadian Forest Service researchers estimate that this figure doubled in 2000.

To make matters worse, the dead timber left standing after these insect outbreaks creates an unprecedented forest fire hazard, the second largest threat (next to insect infestations) to the profitability of the forest industry.

Also, when vast numbers of trees die, so does their potential to reduce greenhouse gases by absorbing carbon dioxide (through photosynthesis) and sequestering it in living wood. Fortunately, there is much research underway in Canada to determine the most effective methods of converting the insect-damaged dead wood into bio-energy (BIOCAP Canada Foundation 2004).

2.   THE NEED FOR PESTICIDES

Managing the overwhelming insect pest problem is a key priority for the preservation of Canadian forests and our forest carbon sinks. This is where pesticides traditionally have come to the rescue. In order to control insect pests both in forestry and agriculture, to improve both timber and crop yields, pesticides of all sorts are regularly used in Canada and throughout the world. Natural, or "biological", pest control products were used in agriculture for centuries until the advent of "improved" chemical pesticides in the 20th century rendered them almost obsolete.

Pest management strategies in forestry have targeted three major groups of pests: pathogenic microorganisms, competing vegetation and insect pests (the latter being by far the most destructive). Historically, pests have been managed by chemical fungicides, herbicides and pesticides. With the subsequent awareness of the negative health and environmental impacts of synthetic pesticide use, there is once again renewed interest in seeking out and commercializing alternative, biological pest controls. Fortunately, the plants and fungi already found in Canada's forests have much to contribute to this effort.

Pesticides are a big business. Agriculture and Agri-food Canada estimates that $190 million is spent in Canada on pesticides annually for food crops alone (Agriculture and Agri-Food Canada 2003). Other estimates of total pesticide expenditures for all purposes (forestry, lawns, golf courses, rights of way, insect control in buildings, etc.) reach $1 billion (Priesnitz 2000). In 1998, herbicides accounted for 49% of world pesticide use, followed by insecticides at 27%, fungicides at 20% and others at 4% (New Internationalist 2000).

At the same time, the World Heath Organization estimates that at least three million people are poisoned by pesticides every year (many of them agricultural workers) and more than 200,000 die (New Internationalist 2000). Farmers using 2,4-D have long been known to suffer elevated rates of some types of cancer as well as reproductive effects (Canadian Environmental Law Association 2000) but until recently, foresters had not been studied. A 2001 study revealed that foresters applying 2,4-D pesticides show genomic instability and elevated levels of certain reproductive hormones (Garry et al. 2001).

Negative economic and environmental impacts from chemical pesticide use also include millions of dollars in public health costs, along with environmental groundwater contamination, honeybee and pollination losses, fisheries, wildlife and bird losses. Arguably, the human losses, health care costs, species decimation and ecological damages associated with the use of synthetic pesticides offset the economic and productivity gains they generate. Not surprisingly, consumer resistance to continued use of chemical pesticides continues to grow; hence, the enormous popularity of organic produce.

Clearly, alternative solutions are needed. In current Canadian forestry practices, conventional chemical insecticides have been removed from the arsenal of pest control weapons because of their serious health and environmental impacts. Thus, there are national government, university and industry research efforts underway aimed at replacing the use of synthetic pesticides in agriculture and forestry with treatments based on the natural (biological) enemies of insect pests and disease pathogens.

3.   BIOLOGICAL PEST CONTROL

Insects, plants and microbes have co-evolved for millions of years. Some insects have developed the ability to digest specific plant compounds while some plants, over time, have built up new defense compounds to repel plant-eating insects or other pests. Scientists are studying these plant defense mechanisms to hopefully mimic them, to then develop commercially useful insecticides and anti-feedant controls.

Forest pest insects, like all organisms, are preyed upon by other animals and suffer from a number of microbial diseases. Among these microbial pathogens are viruses, bacteria, protozoa and fungi. While all microbial pathogens are considered in studies on disease incidence in insect populations, the current Canadian Forest Service research focus for use in microbial control is on viruses, protozoa and fungi (Canadian Forest Service 2004).

Target pest insects being studied are hemlock loopers (*Lambdina fiscellaria*), spruce budworm (*Choristoneura fumiferana*), whitemarked tussock moth (*Orgyia leucostigma*) and balsam fir sawfly (*Neodiprion abietis*). By examining their life cycles, cytology, genetics, epidemiology and methods for mass production, the idea is to determine the impact of naturally-occurring microbial pathogens on insect populations, and the potential of enhancing specific fungi, protozoa and viruses for use as biological control agents.

To this end, the Canadian Biocontrol Network was created and funded by the National Science and Engineering Research Council, as the first multi-stakeholder, "umbrella" organization in Canada to bring together multi-disciplinary, university-based research expertise with relevant partners across the country that share their objectives. Participating organizations include growers, biopesticide and seed companies, government regulators, and environmental monitoring groups. Their ultimate goal is to find alternative solutions to the use of chemical pesticides and to advance biological controls that are safe, environmentally harmless, as well as marketable and socially acceptable.

The Biocontrol Network's research spans a wide range of areas, from the design of "Integrated Pest Management" strategies aimed at reducing the overall amount of pesticide applications in the field, to the function of particular cells in insects and plants, to the use of individual microbes, fungi, bacteria, viruses and small invertebrates as biological control agents specific to targeted pests but harmless to humans and the environment.

Currently, the primary focus of the Network's research program is on greenhouse and tree nursery pests, as a cautious first step out of the laboratory and a prelude to the open field. Tree nurseries are a major economic sector of Canada's forest industry and underpin national reforestation programs. The greenhouse industry is also one of Canada's fastest growing agronomic industries, worth over $2 billion in 2002 (Biocontrol Network 2002).

Pests and disease, still requiring the use of pesticides, afflict both these "contained" ecosystems. Tree nurseries rely heavily on fungicides and broad-spectrum fumigants. Greenhouses use significant quantities of chemical pesticides and fungicides to maintain unblemished produce quality but which, increasingly, are being banned because of toxicity effects.

The Network's approach is to apply varied biocontrol agents in coordinated ways to control insects and diseases, allowing greenhouse and tree nursery growers to progressively eliminate harmful chemical pesticides and still maintain protection of their product. The Biocontrol Network (2002) website provides a list of all microbial and pheromone pest management products registered in Canada, as well as a number of examples of plant defense and mating disruption mechanisms that are now, or are on the verge of becoming, commercially useful insecticides and anti-feedant controls.

It should be noted that the organic farming movement has eliminated the use of pesticides altogether. In the case of severe outbreaks, however, organic farmers and gardeners will use insecticidal soaps mixed with botanical agents such as oil from Neem leaves (a native Asian tree), which doesn't allow many kinds of insects to moult (Howatt 1990).

Despite the vigorous research efforts of the Biocontrol Network, the Canadian Forest Service and others, not many effective, commercially viable methods have been deployed to date. Canadian foresters and farmers have few alternatives at their disposal, and those are mostly based on naturally-occurring *Bacillus thuringiensis* (Bt). Bt is a bacterium that produces a toxin that kills the caterpillar stage of the spruce budworm and other similar moth species. Bt has been the preferred method worldwide to control large insect infestations.

Although the success rate of Bt as a biological control method is acceptable, and it doesn't pose a threat to humans or the environment, its use is sometimes not adequately effective. Bt can be washed out by rain, and sunlight also tends to inactivate the spore and toxins of the bacteria which offsets its positive environmental values.

Biotechnology is a tool that can help identify, select and modify bacteria and other microbials to reduce their limitations and improve their performance, to satisfy the objectives of particular control programs (Globerman and Vertinsky 1995). The Canadian Forest Service is undertaking research to make the mode of action of Bt and other biopesticides more effective (Natural Resources Canada 2005).

Other experts in the field, however, are increasingly concerned that repeated use of Bt and similar substances over a lengthy period will lead to the development of resistance in insects. For example, the diamondback moth (*Plutella xylostella)*, which attacks crops such as cabbage, broccoli, and cauliflower, has already developed resistance to externally applied Bt after repeated spraying in Hawaii, the Philippines and East Asia. With Bt now being so widely used in Canadian forestry practices, and also being enhanced or expressed continually in plants through genetic engineering, it is likely that the rate at which insects develop tolerance will accelerate (Schmidt 1998).

What researchers learn about interactions of biocontrol agents and pest resistance at the contained, more managed level of lab, greenhouse and nursery, certainly will provide valuable models for the open systems of farming, forestry and the more general environment. However, more tools will be necessary for the daunting challenge of protecting trees, crops and other plants while, at the same time, lowering health risks and maintaining a thriving environment.

4.   BIOTECHNOLOGY AND FOREST BIOCONTROL

The Canadian Environmental Protection Act (Canadian Environmental Protection Act) defines biotechnology as, "The application of science and engineering in the direct or indirect use of living organisms, or parts of organisms, in their natural or modified forms." This refers to the techniques through which organisms such as plants, fungi or microorganisms can be used to provide products or services (Canadian Environmental Protection Act 1999). The Canadian Food Inspection Agency (2005) is the federal government agency responsible for the regulation and assessment of plants with novel traits, including forest trees. Biocontrol products, as well as plants with pesticidal properties, are regulated by the Pest Management Regulatory Agency (2005) of Health Canada in accordance with the Pest Control Products Act.

As described above, a variety of parasites, predators and pathogenic microorganisms are being used as natural, environmentally safe biological control agents against forest insect pests, but with various degrees of success. To combat microorganisms causing tree diseases, historically the forest managers have resorted to selecting resistant varieties. There are only a few examples of using fungi for weed control. With the advent of biotechnology, there are now various other options becoming available for protecting trees against not only insect pests, but also organisms that cause diseases and competing vegetation.

Genetic engineering (recombinant DNA technology) has become a common tool of basic research. For over twenty-five years, scientists have been able to modify many living organisms by removing or adding one or more specific genes and by moving genes from one organism to another.

Genes are inherited DNA that carry traits in an organism. For example, physical traits of trees such as size, leaf color and shape, and wood fibre quality are all determined by the information from specific genes. A "transgenic" tree, plant, bacterium or virus is one that has new traits introduced through the use of genetic engineering. With this technique, individual genes from a foreign organism can be removed or replicated from the donor and transferred to a target cell in either a forest tree or a biological control agent. This can result in unique gene combinations with accelerated and commercially improved outcomes unachievable by traditional methods.

Like naturally-occurring biological agents, biotechnology applications are aimed at developing environmentally friendly alternatives to broad-spectrum chemical pesticides. Risk assessment studies are conducted first in the laboratory, then in the greenhouse and, after regulatory approval, in field trials in confined plots. Host plants (trees) can be genetically altered to deter insect or microbial pests and disease, or the biological control agents themselves (Bt) and other host-specific control agents (baculoviruses) can be genetically engineered to both speed up and enhance their effectiveness.

Differing from naturally-occurring bacteria such as Bt, baculoviruses are a group of viruses that also occur naturally, but which are species-specific in their attack of host insects. This makes them excellent candidates for very narrow-spectrum insecticidal applications. The majority of baculoviruses used as biological control agents are in the genus *Nucleopolyhedrovirus* (D'Amico 2003).

When a susceptible insect eats baculovirus, the pathogen particles spread in its tissues and cause disease in the larvae. Consequently, the sick insect stops feeding and dies. However, this process of disabling an insect can take up to two weeks, during which time the insect continues to eat a large quantity of foliage. By genetically modifying the baculovirus (adding or removing a gene), scientists can accelerate its toxic action, greatly speeding up the insect's death, and saving trees from more extensive damage.

Baculoviruses are particularly effective pest control agents as they contain a number of non-essential genes that can be replaced by an introduced gene from another species that is more harmful to the targeted insect pest (Arif 1995). The modification of enzymes, factors and hormones that significantly affect the metabolism, regulate moulting, and generally disrupt normal development of the insect, is being investigated. So far, baculovirus studies have shown no negative impacts on plants, mammals, birds, fish, or even on other, non-target insects. Because they are so much more host-specific than broader bacteria-based toxins like Bt, researchers are attempting to expand the virus host ranges to other pest species as well.

Currently, the Great Lakes Forestry Centre of the Canadian Forest Service is undertaking a large-scale research project co-funded by Genome Canada on the "Genomics of Spruce Budworm and its Viral Pathogens" (Genome Canada 2005). The spruce budworm is by far the most econo-mically important forest pest in Canada, attacking 25% of forest conifers like balsam fir and spruce, the softwood that is used for lumber and pulp and paper. In order to enhance the activity of a baculovirus to control the spruce budworm, it is essential to understand the biology and genomics of the virus as well as its insect host.

The genomics project examines the genes of the spruce budworm and its viruses as well as the genomics of their interaction, with the aim of developing pest control strategies that are environmentally benign. Also, as spruce budworm reproduces only one generation per year, baculovirus resistance is less likely to become a problem. Such research potentially will impact not only the forest, but also the pharmaceutical, veterinary and agricultural sectors as well.

While the safety of recombinant baculoviruses to non-target organisms has been repeatedly shown in the lab, it is essential to demonstrate this in the field before their use is accepted in commercial pest control strategies (Arif 1995). Tests of such genetically engineered baculoviruses have been made by researchers in Canada, Great Britain and the United States and show both safety and promise of effectiveness, although the cost of commercial production of these agents also must be reduced if they are to be competitive (D'Amico 2003).

In addition to studying bacteria and viruses, a deeper understanding of forest ecology can offer further opportunities for the use of both natural and transgenic multi-cellular organisms as pest control mechanisms. Research now focuses (both with micro- and other organisms) on the ability to create a live delivery system of toxins that can be switched off and on, by using enhancers or suppressors external to the organisms. Switch-off mechanisms can prevent the adaptation of insects to biopesticides, thus maintaining their effectiveness for longer time periods (Globerman and Vertinsky 1995). These advances could potentially create a permanent system of pest control with lower environmental risks.

5.   ENVIRONMENTAL IMPACTS OF FOREST BIOCONTROL PRODUCTS

Even within the scientific community, the debate over genetic engineering as an appropriate forest management tool continues. Unlike agriculture, forestry is a production activity carried out in a complex ecosystem that is regarded as "natural". Many people believe that nature should retain the power to regenerate or protect itself and are offended by the view of forests as merely a "resource" or "biofactory". There are ongoing public debates about whether scientists' abilities to transfer genes between organisms through molecular genetics have perhaps outstripped our knowledge of both earth's ecosystems and society's values (Mitchell et al. 1998).

The Government of Canada is committed to ensuring the safety of Canadians and the environment through research, regulations and legislation. To this end, the Canadian Forest Service conducts extensive environmental impact assessments of genetically modified trees and biocontrol products, both to evaluate the potential adverse impacts of these products before they are released into the environment, and to determine if an environmental effect has occurred after product release.

Although, as described above, there are many potential benefits to the use of biotechnology in forest pest management strategies, several hazards have also been identified (Natural Resources Canada 2005a) in introducing genetically modified biocontrol products and trees into the environment. These potential risks include:

- Adverse effects on non-target organisms.
- Gene escape from the genetically modified biocontrol organism or tree to other organisms.
- A change in the interactions within the forest ecosystem.
- Insects developing resistance to biocontrol products.
- Genetically-modified trees becoming invasive and weed-like.
- Impact on genetic diversity and species integrity.

In examining any specific applications of new technologies, such as biocontrol or genetically altering trees and microorganisms, it is important to acknowledge the controversy and recognize that citizens have differing views on the subject. There has yet to be a truly public debate among environmentalists or other members of the public about the relative risks and benefits of genetic engineering, particularly as it relates to the development of a biobased economy, the movement away from the use of fossil fuels and petrochemicals, and the various options for sustainable forest management.

6.   CONCLUSION

Because of the enormous importance of forest values to Canada and to the world, efforts to protect forest biomass from pests are a critical part of forest management. Given the environmental concerns of using chemical pest control methods, coupled with controversy associated with potential biotechnology applications, it is essential that policy makers and researchers alike encourage public awareness and discussion. The ultimate success of the forest bioproduct sector will depend on it.

Chapter 11

OTHER INNOVATIVE PRODUCTS
AND TECHNOLOGIES

Highlights and Fast Facts

⛏	The Government of Canada spends $315 million per year on biotechnology research and development, while industry invests $340 million, and not-for-profit institutes contribute $115 million.
⛏	The natural resources sector of the biotechnology industry in Canada reported revenues of $113 million in 1999.
⛏	Genomics technology is revealing new uses of forest microorganisms for tree augmentation and other environmental applications.
⛏	Bioremediation is an expanding sector that depends on forest fungi, with applications that include agricultural waste reduction, decreasing pollution in watersheds, contaminated sediment cleanup, decontamination of military sites, and bioprocessing techniques in the mining and pulp and paper industries.

1.   USES OF BIOTECHNOLOGY IN FORESTRY

Biotechnology is rapidly transforming many areas of Canadian society. Overall, the federal government spends $315 million a year on biotechnology research and development, industry invests $340 million and not-for profit institutes invest $115 million (Industry Canada 2005a). The natural resources sector of the biotechnology industry in Canada reported revenues of $113 million in 1999. Biotechnology is a fundamental component of the

185

bioproducts sector and will continue to play a pivotal role as the sector grows.

In laboratories across North America, researchers are using detailed knowledge of tree genes and recombinant DNA technology to alter the genetic workings of forest trees, hoping to modify reproductive cycles, growth rate and chemical composition, to change their ability to store carbon, tolerate pests and disease, and to absorb toxins. Much of the research relies on basic tree genetics, made easier by the sequencing of the poplar tree genome, a major effort in forest biotechnology whose results were made public in 2004 (Rosnew 2004).

A complete genome is the ultimate "parts list" for an organism. The increasing availability of these complete genomes is now transforming traditional research approaches in forestry, and in many other scientific fields. The production of transgenic woody plants requires the introduction of a gene into a cell and the growth or regeneration of that cell into an otherwise "normal" plant. The successful integration of gene transfer and tissue culture technologies is, therefore, essential.

In forest tree species, the first introduction and expression of foreign genes was reported for poplar, through a bacterial vector found in *Agrobacterium* (Parsons et al. 1986). Scientists were able to regenerate poplar cells transformed with a bacterial gene that conferred tolerance to the herbicide glyphosate, resulting in the first transgenic trees, and the first gene of commercial value expressed in a woody plant (Fillatti et al. 1987 in Tsang et al. 1995).

The ability to produce transgenic conifer trees in Canada could have a substantial impact on characteristics such as wood supply, wood fibre properties and processing costs. The long generation time of trees (*i.e.*, the time to reach reproductive maturity) makes it difficult to introgress a foreign gene from primary transformants into superior commercial lines through breeding. Genetic transformation could circumvent this problem by allowing the transfer of single gene traits into superior genotypes, leading to the integration of desired traits such as pest tolerance (Natural Resources Canada 2005a).

The success of recombinant DNA technology depends on the ability to cultivate the plant tissues *in vitro*. Somatic embryogenesis (SE) is a tissue culture method for asexual propagation, and is an efficient plant regeneration system that can be scaled up to produce large numbers of tree embryos (Taticek et al. 1991; Beardmore 1995). SE has been an impetus to the understanding of conifer embryo development, and developmental studies are being conducted with the aim of producing vigorous germinants (Roberts and Sutton 1995). Organogenesis, the regeneration of plants by organ formation on various types of tissue and cells, is another *in vitro* method of propagation (reviewed by Chun et al. 1988).

The current emphasis in Canadian tree genetics research is directed towards developing protocols for large-scale production of somatic embryos for commercial forestry. This, potentially, would allow the cost-effective production of planting stock for reforestation. All of the spruce species important to Canadian forestry, as well as Douglas fir and larch, can be propagated through SE (Roberts and Sutton 1995). The Canadian Forest Service has established demonstration plots and field tests of SE-derived trees for white spruce, black spruce and hybrid larch, with the goal of integrating SE into operational reforestation programs (Natural Resources Canada 2005a).

SE offers the potential to add traits through genetic engineering, since propagules result from the regeneration of single cells which are receptive to the introduction of foreign DNA (Roberts and Sutton 1995). Value-added traits, which could be propagated through SE, include yield, wood quality, and pest and disease tolerance. Specific examples where potential applications of SE technology exist are:

- Sitka spruce (*Picea sitchensis*), a high-value species, suffers severe damage from the insect white pine weevil *(Pissodes strobi)*. This damage creates a considerable financial liability for forest companies who are legally required to establish second growth stands. Transgenic trees which are tolerant to weevil could have a significant impact on forest practices.
- Seed supply for western larch is very limited; therefore mass propagation would be a possible solution to this problem.
- Development of a method to select and propagate blister rust tolerance in western white pine (*Pinus monticola*) could benefit this high-value species, which cannot presently be exploited because of extreme sensitivity to this fungal disease.

The motivation to commercialize SE for forestry applications such as tree plantations and ornamental use (*e.g.*, blue spruce) is evident in collaborations between research laboratories and nurseries. These partnerships may be instrumental in producing suitable reforestation planting stock from SE (Roberts and Sutton 1995).

Significant research efforts also are being made in Canada to produce artificial conifer seeds through the encapsulation or desiccation of the somatic embryos (Lulsdorf et al. 1993; Beardmore 1995). The ultimate product would be a widely accessible embryo inoculated with nutrients and pesticides to insure proper growth after storage (Charest and Klimaszewska 1995).

Biotechnology has led to other innovations in the methods of tree germplasm preservation, a tissue culture preservation technique useful for storage of superior tree tissue cultures. These can form part of tree improvement programs, of the production of unique genotypes, and of endangered species (Charest and Klimaszewska 1995).

Cryopreservation (storage at −196ºC) is possibly the best available technology for long-term preservation of tissue cultures, with the benefits of minimal space requirements, and reduced risk of contamination and somaclonal variation (Charest and Klimaszewska 1995). The Canadian Forest Service has developed methods for the cryopreservation of conifer tissue culture lines developed by SE, and has a cryogenic germplasm bank beneficial for long-term storage needs (Natural Resources Canada 2005a).

1.1   Issues in Forest Biotechnology

The research described above is not without controversy. Current issues being addressed relate to gene flow from transgenic trees into natural populations, long-term stability of introduced genes and the potential long-term effects of genetically enhanced trees in the ecosystem (Natural Resources Canada 2005a). Environmentalists and others say that because of the large distances tree pollen can travel, altered genes may indeed migrate to natural populations, leading to damage to ecosystems and other unforeseen consequences.

Tree geneticists are acutely aware that public acceptance will depend at least partly on whether genetically modified trees can be made sterile. Engineered sterility could provide a method for both increasing wood production and genetic containment, and therefore prove to be one of the most important commercial products to be developed for the forest industry using genetic engineering (Regan and Rutledge 1995).

As an example where societal benefit may come of these experiments, Dr. Richard Meagher, a professor of genetics at the University of Georgia, genetically engineered 160 eastern cottonwood (*Populus deltoides*) trees to extract mercury from soil at a contaminated site. Because mercury is an element, it cannot be broken down into harmless substances. Dr. Meagher's "toxic-avenger trees" are intended to remove such heavy metals from contaminated soils in places where other forms of cleanup are prohibitively expensive. The cottonwoods store the mercury without being harmed themselves, convert it to a less toxic form and release the diluted mercury into the atmosphere, where it dissipates and falls back to earth after a few years (Rosnew 2004).

Although this may be perceived as merely moving toxins from one place to another, the risk of human exposure is much lower if the chemicals are not concentrated in certain areas. In time, such trees may be deployed in places like Bangladesh and India, where mercury- and arsenic-laden drinking water has created a growing health crisis.

Forestry researchers also view genetic engineering as a possible way to ease the pressure for logging in wild forests. If they can engineer trees in a plantation setting that grow faster and possess other desirable commercial traits, then the industry will have less incentive to go after old-growth trees.

Among the research goals is the creation of trees that produce less lignin, a molecule that makes the wood fibers stiff. Trees with less lignin can be more readily turned into lumber and paper, using fewer chemicals. Lignin production is obviously important to trees in the wild, contributing to the strength of their trunks, but less so in a plantation where trees will be harvested every few years. Researchers have discovered a link between low lignin and faster growth, which could make the engineered trees particularly desirable for plantation foresters (University of British Columbia 2005).

Scientists are also interested in trying to use genetic engineering to address climate change. Trees can be designed to store (or "sequester") more carbon in their root systems, thereby reducing atmospheric concentrations of carbon dioxide, the heat-trapping greenhouse gas. In a project sponsored by the US Department of Energy, researchers at the Oak Ridge National Laboratory are modifying tree architecture and cell wall chemistry to increase the amount of carbon stored below ground. They are also studying classes of genes that affect absorption of sugars and carbohydrates, which in turn can change the chemical processes that affect the rates at which trees rot and release stored carbon into the atmosphere (Rosnew 2004). Similar research is also being carried out in Canada by the BIOCAP Canada Foundation (BIOCAP Canada Foundation 2004). The Canadian Forest Service conducts extensive environmental impact assessments of all such experiments in biotechnology. It evaluates the potential adverse impacts of genetically modified trees and microbial products long before they are released into the environment, and determines if an environmental effect has occurred after product release.

Nonetheless, it is particularly important for forest researchers, managers and policy makers to acknowledge that, in addition to the considerable benefits potentially offered by this tool, controversy does exist amongst the Canadian public regarding most applications of biotechnology. Before moving experiments from the lab into field-testing, citizens must be consulted about the genetic alteration of trees and microorganisms in the context of their national forest legacy. The same concerns with regard to the use of biotechnology will occur whether the target of the research is for use in the traditional forest industry, or within the emerging bioeconomy.

2.   MICROBIAL PRODUCTS AND TECHNOLOGIES

Microbes, or single-cell organisms, have evolved for some 3.8 billion years and make up most of the Earth's biomass. They have been found in virtually every environment, surviving and thriving in extremes of heat, cold, radiation, pressure, salt, acidity and darkness, often where no other forms of life are found and where the only nutrients come from inorganic matter. It is thought that less than 1% of all microbial species have been described.

In 1994, the US Department of Energy initiated the Microbial Genome Program as a spin-off of its then 8-year-old Human Genome Program. Approximately 100 complete microbial genomes are currently in the public domain for unrestricted use by the scientific community (US Department of Energy 2005).

Microbial research in forestry includes a range of challenges such as soil degradation, environmental waste cleanup and restoration (bioremediation), energy production, biotechnology applications, as well as the biological underpinnings of climate change and the microbial role in the overall processing of carbon and nitrogen on earth (US Department of Energy 2005). The enzymes and metabolic systems of microorganisms have also been found useful in the textile, papermaking and mining industries (Pollution Probe and BIOCAP Canada 2004).

2.1   Microbials and Soil Degradation

Canadian forestry researchers are tackling the problem of soil degradation via inoculation of tree seedlings. Soil quality suffers on certain commercial sites, in the re-establishment of natural forest communities on sites used for other purposes, and for re-vegetation of contaminated or severely disturbed sites (Munson et al. 1995). Specific microbes (bacteria, fungi) can be inoculated into tree seedlings to stimulate their growth in such adverse conditions, proving to be an environmentally preferable alternative to soil fumigation.

Using several different tree species, it has been shown that bacteria present in rhizosphere and/or ectomycorrhizae may stimulate seedling emergence and biomass (Munson et al. 1995). In nurseries, the utilization of Mycorrhization Helper Bacteria (MHB) significantly improved inoculation efficiency by introduced ectomycorrhizal fungi and inhibited infection by competing symbionts (Munson et al. 1995). With these results, MHBs represent an easier and safer alternative to soil fumigation, a pre-requisite to controlled mycorrhization of planting stock in forest nurseries.

2.2   Mycoremediation

Mycelia, the vegetative part of a fungus, can serve as unparalleled biological filters. Mycelium produces extra-cellular enzymes and acids that break down stubborn molecules such as lignin and cellulose, the two primary components of woody plants. These same fungus enzymes that help wood to decompose in nature have proved equally effective at breaking apart hydrocarbons, the base structure common to oils, petroleum products, pesticides, PCBs and many other pollutants.

A major player in the bioremediation industry is Battelle Laboratories, an American non-profit foundation whose scientific goal is to improve environmental health. Battelle is widely used by the US and other governments in finding solutions to toxic waste problems. They have developed a sophisticated process to condition fungal species found in forests to be more efficient at remediating particular compounds.

While Battelle research in mycoremediation is still at the testing stage, contaminants under investigation include petroleum, fertilizers, pesticides, explosives and a wide assortment of agricultural, medical and industrial wastes (Battelle 2005).

Higher wood-degrading fungi are particularly effective in breaking down aromatic pollutants, as well as chlorinated compounds. They also are natural predators and competitors of microorganisms such as bacteria, nematodes and rotifers. Proprietary strains have been developed that detect, attack and destroy or inhibit the growth of bacterial contaminants, such as *E. coli* (Battelle 2005).

Classified research is also being conducted by the US Defense Department on fungal strains that can be used in the destruction of biological and chemical warfare agents. A particular mushroom species has been found to break down agent VX, a potent nerve gas employed in Iraqi missiles during the 1990s Gulf War that is very difficult to destroy (Mycova[SM] 2005).

2.3   Mycorestoration

Forest harvesting and other developments have caused rural roadways to become a primary source of silt runoff and pollution to watersheds and sensitive ecosystems. The building of roads and the resulting compaction creates an environment absent in mycorrhizal fungi (Stamets 2005). This hinders the recovery of native flora and habitat.

Mycorestoration, another industrial use of mycelia, is emerging as a way to deal with this problem on, for example, abandoned logging roads (Stamets 2005). The mycorestoration process involves placing hog-fuel (bark and wood chips) into logging roads. This wood debris is then inoculated with mycelia of a mosaic of keystone native fungal species. The fungified wood chips prevent silt-flow through the natural filtration properties of the mycelial networks, and in the process renew topsoils, spurring the growth of native flora and fauna.

The novelty of mycofiltration is the purposeful introduction of fungi to the wood chip buffers, which enhances and accelerates the decomposition essential for habitat evolution. This method deters water contamination and jump-starts the process of recovery, allowing nature to steer the course of species succession.

Microbes found in the forest may provide relatively simple solutions to a complex field of problems. When the full costs are taken into consideration ecologically (forests, watersheds, fish ecology), economically (lumber, road construction, access), and aesthetically (other forest values, recreational uses), these various microbial products and technologies are worthy of serious consideration.

3.   INNOVATION IN THE PULP AND PAPER INDUSTRY

Forest-related biotechnology also is poised to offer potential solutions to health and environmental problems associated with the production of pulp and paper. Through genomic and bioremediation breakthroughs, research on the combination of tree improvement and the modification of micro-organisms for bioprocessing is likely to give rise to many types of novel products (Globerman and Vertinsky 1995).

Traditionally, pulp is made by boiling wood chips in a chemical solution to break down the lignin and fortified cell walls of the wood. Biopulping is defined as the treatment of wood chips with lignin-degrading fungi prior to pulping (US Department of Agriculture 1996). Some pulp and paper companies are looking at using such fungi (for example, white rot, or *Bjerkandera*, which plays a key role in the natural rotting of fallen trees), as well as fungal enzymes, to begin the breakdown of lignin and fibre separation of wood prior to mechanical pulping (Pollution Probe and BIOCAP Canada 2004).

The fungal pre-treatment of wood chips prior to pulping reduces the amounts of electrical energy and chemicals required during refining. It also increases mill throughput, improves paper strength and limits the amount of waste, resulting in direct economic and environmental benefits (Scott et al. 1998). For example, European paper producer Leykam Austria has found that, as a pre-treatment for pulping, the fungal enzymes contribute to the removal of 30% more lignin with the use of less chlorine bleach (Pollution Probe and BIOCAP Canada 2004).

An economic analysis of a 600-t/d thermomechanical pulp mill indicated that based on energy savings, the process results in savings of about US$10 per ton of pulp. As well, increasing the mill throughput by 20% resulted in additional savings of over US$40 per ton of pulp (US Department of Agriculture 1996).

Biotechnologies in various stages of the research, development and commercialization process include the following (Bourbonnais et al. 1991 in Globerman and Vertinsky 1995):

- Biomechanical pulping: pretreatment of wood chips for mechanical pulp with various fungi has been shown in the laboratory to reduce refining energy and improve strength.
- Liginase and biomimetic pulping: the discovery of ligninperoxidase (liginase) in 1983 led to hopes that the enzyme would be applied in pulping and bleaching. The original reports claimed that the enzyme was capable of depolymerizing certain lignins. Pulping and bleaching experiments are ongoing.
- Wood protection during chip storage: a fungus was identified as being capable of protecting wood against rot in laboratory tests, making this potential application an open opportunity.
- Control of slime: glucose via enzymic hydrolysis of primary clarifier sludge in pulp and paper operations.
- Color removal from pulp bleaching effluent: the interest in treating bleachery effluents primarily focuses on the removal of toxic chlorinated organics, rather than color removal.

4. CONCLUSION

The Organization for Economic Co-operation and Development has defined innovation as the process through which new economic and social benefits are extracted from knowledge. If managed responsibly and transparently, there is no doubt that innovative products and tools, such as those created though biotechnology, have the capacity to revitalize the forest-based bioeconomy. As well, new jobs and business opportunities, in both resource-based communities and knowledge-based industries, can be created, while at the same time lessening adverse impacts on human health and the environment (Pollution Probe and BIOCAP Canada 2004). Such novel technologies also have the potential to increase Canada's international competitiveness and promote sustainable development in the nation's key economic sector – our forests (Industry Canada 2005a).

Chapter 12

CARBON CREDITS

Highlights and Fast Facts

- Propagation of trees with "preferred" traits such as disease and insect tolerance, drought tolerance, cold-hardiness, reduced lignin and increased biomass, can provide needed feedstocks for bioenergy production, platform biochemicals, plastics etc., while at the same time alleviate pressures on old-growth and recreational.
- The Kyoto Protocol on Climate Change, an unprecedented treaty in international law, came into effect in February 2005, establishing legally binding targets for reducing greenhouse gas (GHG) emissions in developed countries. It also instituted an emissions trading (ET) and carbon credit system, potentially impacting on bioproduct development and forest carbon management (FCM) practices in Canada.
- The opportunity for FCM projects and carbon credit trading for Canada is significant, potentially worth an estimated $60 million, but owing to the complexity of the Kyoto Protocol not all forest carbon can be accounted as GHG offsets.
- There is a potential double benefit to the recognition of forests as carbon reservoirs as they can be valued for their contribution to mitigating climate change while, at the same time, old-growth, regeneration, wilderness and biodiversity are preserved.
- A number of policy refinements are required to facilitate the ET system in the immediate future. In addition, the use of harvested wood products and bioproducts as a permanent source of carbon sequestration and GHG offsets will no doubt be a key issue of negotiation in the next Kyoto Commitment Period (post 2012).

1.   THE KYOTO PROTOCOL

In this chapter we explore the constraints associated with using carbon credits as a means to generate additional revenues from forests. From an intuitive perspective a growing forest can accumulate carbon, which in turn should be used in carbon crediting. In practice, however, carbon accounting is a complex issue with biological, political and contractual constraints.

On February 16, 2005 the world's first legally binding international treaty on the environment, the Kyoto Protocol, came into effect. Thirteen years after the Climate Change Convention was agreed to at the Rio Earth Summit in 1992, eight years after each nation's targets for cuts in greenhouse gas (GHG) emissions were defined in Kyoto, Japan in 1997, and three years after Canada ratified the Protocol in 2002, a complex system of auditing GHG emissions for each country was at last formalized.

The Kyoto Protocol, with its accompanying emissions trading (ET) and carbon credit system, is unprecedented in international law. Persuading countries to surrender their national sovereignty over domestic industrial policy was so controversial, yet so vital to addressing climate change that greater GHG reductions were left until the next round of international negotiations.

Some countries such as the United States (which alone accounts for 36% of the industrial world's emissions) did not ratify the agreement. Nor did Australia, Monaco or Liechtenstein take part. In addition, developing countries, including the rapidly industrializing China and India, were exempted from the first round of cuts, the rationale being that our existing GHG problem was caused almost entirely by industrialized countries.

Carbon dioxide (CO_2), one of the six primary GHGs, accounts for over 80% of the total GHG emissions from developed countries, and is largely generated by fossil fuel combustion and deforestation—roughly 20 percent of global emissions actually are caused by deforestation (Ross 2004). Because the impact of each GHG on climate change varies, the emissions of each gas are translated into CO_2 equivalents (CO_2e) based on their global warming potential. For instance, the global warming potential of methane is 11 times that of CO_2 so methane would have 11 CO_2e/molecule while CO_2 would have only 1.

The Kyoto Protocol came into action when the required minimum of 55 countries, (including developing countries), accounting for at least 55% of developed countries' 1990 carbon dioxide (CO_2) emissions, ratified the agreement. It now commits developed countries to reducing their collective GHG emissions by 5% below 1990 levels. Canada's allocation is 6% below 1990 levels. In 2005 this actually means reducing national emissions by closer to 30% from current levels, because we are emitting so much more now than we were in 1990 (NRTEE 2002).

Current agreed-upon reductions are to be completed between 2008 and 2012, known as the "first commitment period". The average annual emissions during this time-frame will be used to determine whether individual countries have met their reduction target. Negotiations for the second commitment period targets and rules are planned for the latter part of 2005.

When the Protocol was first negotiated in 1997, the most innovative element may have been the three kinds of international exchanges of carbon credits it provided for. Industrialized countries and the business community pushed hard for the so-called "flexibility mechanisms", which they saw as key to reducing the costs of complying with the Protocol's targets. These instruments were also regarded as important to the Protocol's sustainable development goals, by helping fund emissions reduction projects in developing countries, by enabling the sharing of emissions-reducing technologies, and by permitting Emissions Trading (ET).

Under the Protocol, individual countries can meet their targets through various domestic policies and economic measures to reduce their emissions and potentially generate carbon credits. These can include curtailing emissions at their source (tail pipes and smoke stacks); making efficiencies to consume less energy; using alternative energy sources that emit fewer GHGs (including biofuels); or offsetting emissions through carbon sequestration practices. These latter methods can include the management of "carbon sinks" such as geological (mines and oil shafts) and biological (oceans, forests, agricultural fields, wetlands and soils) to absorb, or sequester, CO_2.

The concept of ET is still not widely understood. Under the Kyoto protocol, countries can engage in emissions trading as a way of meeting their international obligations. This is distinct from emissions trading domestically. In an effort to meet international obligations, individual countries issue permits to firms and industry to emit a certain amount of emissions over a certain amount of time. Firms can then trade these credits. Domestic trading is based on the idea that different sectors of the economy will have different costs associated with reducing GHG emissions. By generating "carbon credits" that can be bought and sold in the market place, all companies can meet GHG reduction obligations set by governments more readily. For instance, if Industry A is able to cut its emissions at low cost, it also has a financial incentive to make extra reductions because it will be able to sell the surplus as carbon credits. Industry B, which may face much higher costs to reduce its own emissions, will buy the credits from Industry A as it can save money but still meet its reduction targets. The overall outcome is that the aggregate emissions reductions are achieved (*i.e.,* the atmosphere benefits) but at a lower overall cost. This trading tool thus, theoretically, will help minimize economic disruption and protect competitiveness in the transition to a carbon-constrained world (NRTEE 2002).

Concentrations of CO_2 in the atmosphere have reached an all-time high of 378 parts per million (BIOCAP 2005). Measured against the potentially catastrophic consequences of global climate change, the trading mechanisms and modest controls on GHG emissions to be enforced by the Kyoto Protocol seem trivial. It is some comfort, at least, that the principle has now been established in international law, and that the next round of post-2012 targets and rules will likely involve much deeper cuts to GHG emissions. The developing nations currently exempted from Kyoto controls will no doubt have to accept stringent emissions reductions targets in future as well.

1.1   Greenhouse Gas Emissions and the Biosphere

Past experience and current research strongly suggests that maintaining and enhancing forest carbon reservoirs is among the least costly and immediately available options for offsetting CO_2 emissions (Totten 1999). In fact, if it wasn't for our planet's biosphere, our climate change problem would be far worse than it is today. The earth's terrestrial systems are thought to absorb approximately 30% of the CO_2 added to the atmosphere from human-induced activities (Houghton et al 2001).

It is likely also possible to increase the rate at which ecosystems remove CO_2 from the atmosphere and store the carbon in plant material, including trees, and in decomposing detritus and organic soil. In essence, forests and other highly productive ecosystems can become efficient "biological scrubbers" by removing, or sequestering, CO_2 from the atmosphere.

The fact that policy makers are giving serious attention to forest carbon sequestration and offset programs can partly be explained by claims that this approach provides a relatively inexpensive means of addressing climate change, especially compared to other GHG reduction options. Canada's massive forest cover appears to give it a huge natural advantage in this respect, although some uncertainty exists owing to biological uncertainties.

Estimated costs for sequestering up to 500 million tonnes of carbon per year—an amount that would offset up to one-third of current annual US carbon emissions—range from US$30 to $90 per tonne (Stavins and Richards 2005). On a per-tonne basis, these costs are comparable to those estimated for other climate change mitigation options such as fuel switching or energy efficiency.

Although the biosphere certainly helps to lessen the impact of fossil fuel emissions on atmospheric concentrations of GHGs, there is a very large year-to-year variability in the contribution of forests and soils in carbon sequestration. Forests are large reservoirs of carbon accumulated in trees and soils. In order to build up carbon, growing forests extract CO_2 from the atmosphere through photosynthesis. However, this store of carbon can be released back into the atmosphere if forests are harvested, or under attack by insects, disease or decay, or burn in fires. In these cases, forests can become a "source" of GHGs.

These key sources of uncertainty, mainly fires and insect infestations prior to and during the Kyoto commitment periods, are both unpredictable and not truly manageable. This unpredictability is a critical challenge to understanding the contribution of forest sinks to the Kyoto Protocol, especially in Canada, since the nation's biological carbon stocks—especially soil carbon—are among the largest in the world (BIOCAP 2004a). Therefore, an intensive university-government-industry research program is underway to understand variables affecting carbon cycling in Canada's forests. Scientists in the Fluxnet-Canada Research Network are studying how various biological systems respond to changing climate regimes, including temperature, season length, drought, wind, fire and pests. Emphasis is on measuring, computer modeling and analyzing the movement of CO_2 between the atmosphere and various forest and peatland ecosystems across Canada in an attempt to understand the nature of terrestrial carbon sources and sinks (BIOCAP 2004a).

Sustainable forest management (FM) practices are also being closely examined. Strategies such as reforestation, control and reduction of losses due to fires, harvesting, insects, disease and decay, can reduce the role of forests as a source of GHGs. If FM policies, practices and technologies can also increase the rate at which carbon is absorbed in growing forests, so that it is greater than the carbon released due to fires, harvest, decay, insects and disease, then forests can be considered an important sink for CO_2 which in turn may fall under the offset system under the Kyoto Protocol. It is important to understand that not all carbon accumulated by forests is accepted as creditable under the Kyoto Protocol.

1.2   Forest Carbon Management and the Kyoto Protocol

Canada is home to about 10% of the world's forest cover and 25% of the world's natural forests (Lazar 2005) and these percentages are likely to increase with further global deforestation and plantation forestry. The Canadian position throughout the prolonged international negotiations associated with the Kyoto Protocol has been that emission reductions generated by sound forest management practices should be permitted and even encouraged under the Kyoto Protocol. Including forests under the Kyoto Protocol might provide financial and political incentives to enhance and protect sinks and reservoirs while at the same time promoting sustainable forest management. Hence, the agreement does contain a variety of rules to allow and regulate how such reductions can be counted as part of a country's emission cutback scheme.

Specifically, Articles 3.3 and 3.4 of the Protocol require countries to account for a range of FCM activities, including what is now known as "ARD": Afforestation/Reforestation/Deforestation. They also allow FM activities that either enhance carbon sequestration, or reduce and avoid GHG emissions. For the purpose of carbon accounting, how to count a particular landscape as a carbon sink or source is described by specific rules.

Each country that decides to account for FM under Article 3.4 is required to define its "Kyoto Forest". This is the area subject to forest management for which a nation will account for GHG emissions and removals in the first commitment period. Countries have until August 2006 to decide whether they will use this forest, and/or agricultural cropland and grazing land management, and/or re-vegetation activities, to help meet their emissions targets. At the end of the first commitment period in 2012, each nation must account for changes in carbon stock and either add it to its Kyoto Protocol targets if it turns out to be a net source, or subtract it from the target if it is a net sink. Considerable efforts are underway to develop the forest carbon measurement and reporting system to enable Canada to account for carbon stock changes under both Article 3.3 and 3.4 (Canadian Forest Service 2005a).

The international carbon-trading "flexibility mechanisms" have direct implications for global forests. The Clean Development Mechanism (CDM) allows developed countries to acquire credits through GHG emissions reductions projects (including bioenergy projects) as well as afforestation and reforestation projects in developing countries. Joint Implementation (JI) projects allow developed countries to invest in each other's sinks enhancement projects to gain credits. Emissions Trading (ET) lets developed countries trade permits that allow them to emit certain levels of CO_2e. These "allowances" are expected to be traded under the Kyoto Protocol's "cap and trade" system where each developed country is allocated a maximum allowance which is equal to their 1990 emissions multiplied by their target reduction percentage (*e.g.*, 94% for Canada). Just like companies, countries that reduce their emissions below their allowance through efficiency improvements or sink enhancement activities will be able to trade some part of the surplus allowance to other developed countries (United Nations Framework Convention on Climate Change 2003).

In the Kyoto Protocol, with forest sink enhancement accounting based on the national inventory reports, the annual increase in the size of a forest carbon sink is represented by the "removal unit" (RMU). Under the international carbon-trading rules, unused credits can usually be saved, or banked, from the first to the second commitment period (United Nations Framework Convention on Climate Change 2003a). The exception to the rule is the case of RMUs, which are not bankable, due to the fact that forests can change drastically in a short time. Indeed, forests can switch from accumulating carbon as sinks during growth periods, to net releasing carbon as sources in result of damage or destruction through unpredictable catastrophic disturbance, such as forest fires or insect infestations.

Participating countries must account for emissions and removals from ARD activities and, currently, there is no cap on these activities. However, if ARD activities turn out to be a net source of emissions, the extent to which these can be offset in Canada through FM activities is limited to 33 million tonnes of CO_2e emissions per year during the first commitment period (2008-2112). Canada also can use domestic FCM activities and JI projects involving FM to provide an additional sink offset of up to 44 $MtCO_2$e/yr (United Nations Framework Convention on Climate Change 2003a).

The use of forests as carbon sinks to generate offsets isn't without controversy. Indeed, many environmentalists initially opposed the inclusion of forests as carbon sinks in the Kyoto Protocol, until it was understood that such provisions could promote forest conservation (Ross 2004). As such the Kyoto Protocol can be used to favour longer rotations, selective logging and reduced harvesting levels, which are all practices advocated by forest conservationists. Better yet, the revenue losses normally associated with these practices may be partially offset by the carbon saved, if considered a credit (Ross 2004).

The Pew Centre for Global Climate Change (Stavins and Richards 2005) has identified several key factors that affect estimates of the cost (and therefore the practice) of forest carbon sequestration:

1. The tree species involved, forestry practices used, and related rates of carbon uptake over time—growth rate is affected by tree species;
2. The opportunity cost of the land—that is, the value of the affected land for alternative uses
3. The disposition of biomass through burning, harvesting, and forest product sinks
4. Anticipated changes in the pricing of forest and agricultural commodities
5. The analytical methods used to account for carbon flows over time
6. The discount rate employed in the analysis
7. The policy instruments used to achieve a given sequestration target

Although use of forest sinks, land management practices and carbon storage are not a total solution to reducing GHG concentrations in the atmosphere, they are still very important to the overall global "carbon budget". Thus, much effort is being undertaken worldwide to understand and reduce uncertainties about carbon dynamics. Accurately measuring carbon level changes certainly is critical to establishing fair and cost-effective international ET policies.

Whether and how to include biological sinks and sources will no doubt continue to be a point of contention for the post-2012 negotiations amongst environmentalists, scientists, the forest industry and politicians alike. The factoring out of natural and indirect effects from the direct effects of FM activities, as well as the use of harvested wood products as a permanent source of carbon sequestration, will be key issues of debate (Booth 2005). The potentially significant role of forest bioproducts that may store carbon, or bioenergy that reduces or avoids GHG emissions, therefore will also be carefully considered.

1.3   Forest Carbon Management and Emissions Trading

Canada's 2005 "Kyoto Action Plan" (Government of Canada 2005a) places substantial, although certainly not exclusive, reliance on market mechanisms. For example, Canada established a $1 billion "Climate Fund" which goes in part to funding international credit purchases. It has also proposed a set of rules for a domestic offset credit system, which is intended to "reward innovation and provide incentives to reduce GHGs" (Environment Canada 2005). Under such a scheme, land owners engaging in state-of-the-art FCM practices may benefit.

Undoubtedly, assessing the financial outputs of carbon credits entails a number of variables that can be affected by a wide array of decisions and policies that are still under debate at the national and international levels. However, within the context of this book it is important to understand the potential contribution of carbon credits to the economics of silviculture, as it may provide additional value to forest ecosystems and is the object of speculation.

The price of CO_2e emission allowances on the European Union's Emissions Trading Scheme already reached a record Cdn. $31 per tonne (BIOCAP 2005a). However, it is important to note that the EU system is unique to the EU and does not trade in Kyoto compliance units. As EU units are not valid in any other system, Kyoto or Canadian, Canada's system will be only trading in units valid in Canada and the main buyers will be Large Final Emitters and the Climate Fund. As both the Offset System and the LFE system are still in development, it is difficult to estimate future offset credit prices, however expectations are for below $15/tonne.

Canada's preliminary estimate of the contribution that FCM will make to its Kyoto commitments is that forests could sequester 20 Mt CO_2e per year from 2008-2012. This figure is determined by projecting the net impacts of all "business as usual" forest-based activities (ARD and forest management) eligible under the Protocol on lands likely to be selected for national reporting (Canada's "Kyoto Forest"). Work is underway to improve our understanding of the potential contribution of FCM, and the probability of different outcomes (Booth 2005). In reality, what Canada will be able to claim is the actual carbon stock changes that take place during the commitment period, which may be more or less than this estimate.

Currently, the federal government estimates that FCM projects could generate more than 4 Mt CO_2e/yr by 2010. Calculated at $15/tonne, the carbon credits created by these projects would be worth over $60 million, thus representing a very significant economic opportunity.

Perhaps more importantly; sustainable FCM projects could contribute significant environmental and social benefits. Although the extent to which the following would be creditable under an offset trading system is unclear, forestry practices that increase carbon sequestration include (Stavins and Richards 2005).

1. Afforestation of agricultural land
2. Reforestation of harvested or burned timberland storage, which are emerging practices
3. Modification of FM practices to emphasize carbon storage
4. Adoption of low impact harvesting methods to decrease carbon release
5. Lengthening forest rotation cycles
6. Preservation of forestland from conversion
7. Adoption of agroforestry practices
8. Establishment of short-rotation woody biomass plantations
9. Urban forestry practices

In general, the objective of an FCM project is to reduce net emissions of carbon to the atmosphere from the project site (*e.g.,* by reducing deforestation) or to increase carbon removals from the atmosphere on the project site (*e.g.,* by establishing fast-growing plantations), in comparison to what would have happened in the absence of any project. The result of these activities is a carbon credit, or offset, which may be retained by the project proponent or traded in the Canadian offset trading system.

Overall, some of the benefits of FCM projects are as follows (Griss 2002, 2004):

1. Many of the policies and practices needed for carbon management also serve other forest management objectives. These include: more efficient and sustainable use of forests, increased biodiversity and reduced environmental impacts from forest harvesting
2. The cost per tonne of carbon absorbed may be significantly lower than the costs of other emissions reductions strategies
3. Reductions from FCM might be integrated with an ET system that would provide economic incentives to more sustainable forest management practices
4. New economic opportunities could arise, including: cogeneration, bioenergy and other forest bioproducts, agro-forestry, manufacturing using wood wastes, and export of technologies and expertise in forest management and bioproduct development

Some difficulties with FCM to generate emissions reductions are as follows (Griss 2002, 2004):

1. Measurement and verification of reductions may be difficult or costly, *i.e.*, the issue of additionality
2. There are natural limits to how much and how long carbon can be sequestered in forests, *i.e.*, the issue of permanence
3. Carbon accumulation in forests can be a slow process and may not have a major impact in time to count towards Kyoto targets
4. Natural destructive forces leading to large releases of CO_2 such as fire, insects and disease may be too difficult or costly to control, leading to unforeseen emissions that will have to be compensated by other more costly reduction strategies

2.   CURRENT TRADE IN CARBON CREDITS

To date, the lack of clear ET rules world-wide certainly has hampered the development of a global GHG carbon credit market. Nevertheless, some governments, including Canada, have moved towards establishing domestic trading programs as in the United Kingdom, Denmark and the European Union, while Norway, the Netherlands and Australia are examining their feasibility (Rosenzweig et al. 2002). A total of 284 transactions involving 335,000,000 tCO_2e were recorded from 1996 to September 2002, approximately 45% of which occurred in 2002 (Carbon Market Analyst Online 2002). Prices have ranged from $0.83 to $12.76/tonne. Most of these trades have occurred under unregulated, voluntary frameworks involving commodities defined by the trading participants, and not within government-defined trading programs. These are commonly known as verified emissions reductions (VERs) and only contain the possibility, and not the guarantee, that governments will treat them as allowances or credits that can be applied against future emissions reductions requirements (Rosenzweig et al. 2002). Although each nation officially controls the total number of allowances or credits, the compliance requirements, how permits/credits are allocated, and the degree to which they can be transferred within their domestic trading systems (IETA 2002), individual transactions are, in practice, currently controlled by industry. Nonetheless, Government-issued permits are expected, in future, to have a significantly higher value than VERs. Past estimates indicate that carbon prices could reach US $33 to $44 per tonne in the US and US $77 to $88 per tonne in European and Japanese markets when full trading exists under the Kyoto Protocol (Totten 1999).

Carbon credit buyers include large oil and gas companies, electric utilities and other firms that emit significant volumes of GHGs and anticipate future emissions limits. In 2001, 33 companies with US assets in the mid-west formed the Chicago Climate Exchange (CCX) and planned to explore the potential for a regional GHG trading exchange that would give credits for domestic as well as international sinks projects. The exchange is expected to expand its operations to the entire US, Mexico and Canada (Rosenzweig et al. 2002). As a way of further legitimizing such transactions, the World Bank established a BioCarbon Fund for projects that sequester GHGs in forests, agricultural areas and other ecosystems (Ross 2004).

2.1   Impact on Forests

Carbon buyers face a number of choices when developing their carbon offset portfolios. They need to decide whether to purchase sinks or emissions reductions offsets. If sinks, do they prefer offsets from forest or agricultural sinks? If forest sinks, will they be domestic or international? Are the sinks compliant under the Kyoto Protocol? Finally, if international, do they prefer projects in developed or developing countries? Domestic forest carbon offsets therefore, need to be very competitive with other domestic and international sources.

Although forest carbon sink activities are considered a cost-effective means of reducing GHGs, due to uncertainty regarding future treatment of carbon sequestration under the Kyoto Protocol, the demand for forest carbon sink offsets has been limited. Moreover, Canadian GHG emitters have found domestic forest carbon management projects to be less competitive than international projects due to cheaper land and faster tree growth rates in foreign jurisdictions (Griss 2002).

The Climate Trust, an Oregon state-sanctioned nonprofit entity that secures carbon offsets, provides an example of the impact of carbon trading on the forest sector worldwide. Power plant developers are required by the State of Oregon to offset emissions that exceed a specified output. Plant developers can choose to purchase qualifying assets in the market or pay $1.27 per tonne of CO_2 to The Climate Trust. To date, all developers have chosen the latter path.

As a result, the Climate Trust is funding five offsetting projects from the first US $1,500,000 in payments and a cash position of $8,250,000 from its second round of payments. One of the first projects involved paying the Lummi Indian Tribe to preserve 364ha of old-growth Pacific forest as a sink. This undertaking should yield least 350,000 tonnes of CO_2 offsets over 100 years. As the United States has not ratified the Kyoto Protocol these carbon offsets are tradable with Kyoto Protocol units, which shows the presence of a disconnect between on the ground activities and compliance to the Kyoto Protocol.

In Canada in the early 2000's, exploratory emissions trading programs such as CleanAir Canada's pilot emissions reduction trading program and the GHG Emissions Reduction Trading (GERT) pilot served to familiarize Canadian companies with emissions trading (Rosenzweig et al. 2002), and this regardless of compliance to the Kyoto Protocol. While it is difficult to ascertain the precise level of trading activity within Canada, as of early 2002, the CleanAir Canada program had evaluated approximately 50 projects involving more than 43,000,000 tonnes CO_2e in emissions reductions, 5% of which involved biological sequestration (CleanAir Canada 2003).

Certainly there are several Canadian companies, in particular, Ontario Power Generation (OPG), TransAlta, Epcor and Suncor, who have been among the most active buyers in the carbon market and tend to prefer purchasing at least some of their reductions from local FCM projects.

Alberta-Pacific Forest Industry Inc. (Al-Pac), Saskatchewan Environment and Resource Management (SERM), the Tree Canada Foundation (TCF) and OPG provide other recent examples of the impact of ARD activities and forest carbon offset trading by industry, government, NGO and a Crown Corporation on the Canadian forest sector (Griss 2002). Al-Pac leases cleared land from farmers in northern Alberta, plants fast-growing poplar trees, and pays farmers a maintenance fee to look after the plantations. In addition to providing a source of timber, Al-Pac predicts this will help offset its emissions and sequester 22,400 tCO_2e per year by 2006 and generate 455,000 tCO_2e per year of credits by 2024.

Al-Pac also encourages farmers who are considering to clear their forested land, to manage their timber sustainably. This provides farmers with an additional and ongoing source of income, ensures a timber supply for Al-Pac, as well as maintenance of forested land, and avoids or delays potential deforestation (Climate Change Central 2002a).

SERM is planting 3,300 ha of land with white spruce trees to sequester carbon. An additional 200,000 ha of forest will also be removed from the annual allowable cut to avoid forest management-related emissions as part of Saskatchewan's Representative Areas program. In addition to the emissions avoided, the project is predicted to absorb 115,000 tCO$_2$e over a 90 year rotation. The earned credits are, in turn, sold to SaskPower.

TCF planted 63.3 million trees over a period of six years in a partnership with a range of agencies and communities across Canada that started in 1997. These have been offered as credits to notable sponsors such as TransCanada Pipelines.

In another project that combines forest carbon sequestration with conservation of biodiversity, by the end of 2005, OPG will have planted 1.6 million native trees and shrubs in southern Ontario and an additional 200,000 trees per year thereafter to offset hydro emissions. This project is focused on regional forest habitat restoration through expansion of core forests and establishment of habitat corridors.

Regardless of whether the above FCM activities are considered eligible under Canada's offset trading rules, such activities already have shown to provide positive conservation benefits through the creation of shelterbelts and urban forests; restoration of forests; establishment of conservation easements and creation of plantations to replace fossil fuels and reduce logging in non-plantation forests. FCM activities can also provide positive benefits by encouraging lengthened rotation age, management of old growth forests, selection logging and establishment of large protected areas.

2.2   Implications for Policy

Carbon offsets presently can be generated at low cost, add financial and ecological value, and diversify existing investments. Most importantly, the scientific and business expertise exists in Canada to capitalize on FCM projects. However, investors will continue to avoid the political and natural risk, the lack of clarity around ownership of credits, and the complex measurement and verification activities associated with FCM projects until these issues are squarely addressed. For instance, ARD rules have yet to be defined for the second commitment period under the Protocol, so there is a risk that the above ARD activities eventually may be considered ineligible or discounted in the future. Canada's Kyoto Forest also still needs to be clearly specified.

Another contentious issue is that, while timber harvested for either wood or paper products will likely be treated as carbon releases for the first commitment period, a decision must be made about both wood and bioproducts in future commitment periods; *i.e.,* should they be treated more like a sink which retains carbon in the longer term? For instance, dimensional lumber and other non-timber forest products can fix carbon as long or even longer than they would had the tree not been harvested (as opposed to paper, which degrades far more rapidly).

Other tricky policy matters include the claim from provinces and territories that the benefits from forest sink assets must accrue to the province or territory that 'owns' the assets. Since the majority of Canada's forests (crown lands) are under provincial or territorial control, with only 5% under direct federal control, extensive cooperation with provincial and territorial governments obviously will be required.

A further distinction needs to be made about precisely who owns carbon credits for the purposes of trading. While increases in forest carbon sequestration can be reported internationally as RMUs regardless of who is responsible for securing the increase, decisions must be made about which entities will secure the carbon credits that derive from incremental increases. They would then also be liable for incremental decreases from business-as-usual activities (Griss 2004). Some public lands are directly managed on behalf of the Crown by forest companies so the actions that can be taken will vary. Private lands may or may not be used to produce forest products, and urban or municipal forests may also have a role to play in FCM. The Federal government launched a feasibility study, Feasibility Assessment of Afforestation for Carbon Sequestration (FAACS), to determine how private landowners can be encouraged to engage in large-scale afforestation activities (CFS 2005). A cost-benefit information system for afforestation in Canada has been developed (McKenney 2004), which simulates a wide range of possible benefits and costs associated with afforestation.

The Forest 2020 Plantation Demonstration and Assessment initiative was launched in 2003 (CFS 2005) to explore policy options related to the potential role for afforestation to sequester carbon and thus partially meet Canada's Kyoto targets. A major component of the initiative, set to wrap up in March 2006, is the establishment of some 6,000 ha of forests on underutilized agricultural lands across Canada.

3.   CONCLUSION

Canada has the potential to become a world leader in ARD, FCM and forest bioproduct activities given the extent of its natural resources base.

While policy refinements urgently need to be made, evidence indicates that forest companies, landowners and local governments have been ready to move for a while (Amey 2003).

Some of these policy imperatives include the need for:

1. A standardized model, currently under development, including baselines, for calculating the flow of carbon and carbon stock changes within a forest, to ensure reporting according to international guidelines (CFS 2005)

2. Improving the precision of forest carbon sequestration estimates by developing better scientific methods to track, measure and monitor carbon absorption and release activities (BIOCAP 2004a)

3. Reasonable certainty for buyers, sellers, regulators, and the general public, that carbon credits represent valid emissions reductions (Pollution Probe 2003)

4. Encouragement of public policies and markets that recognize the carbon sequestration benefits of forest products (Miner 2005)

While both domestic and international emissions trading are challenging to implement, when properly structured, they make great environmental and economic sense for dealing with GHGs.

Chapter 13

THE WAY FORWARD

Highlights and Fast Facts

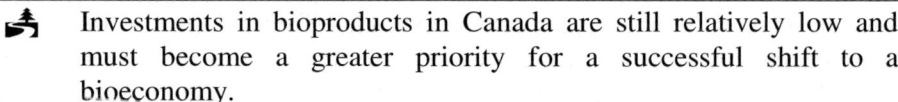 Investments in bioproducts in Canada are still relatively low and must become a greater priority for a successful shift to a bioeconomy.

Investments in bioproducts in Canada are still relatively low and must become a greater priority for a successful shift to a bioeconomy.

A commitment to research, development and commercialization is needed for Canada to become a leader in this new sector and benefit from our "green" advantage.

A bio-based economy will require researchers cross-trained in diverse disciplines who can bridge the gap between the academic sector and the industrial sector.

Public debate about environmental and ethical concerns of the impact of forest bioproducts will be crucial to the development of the bioproducts industry and the corresponding regulatory framework.

As the crow flies from St-John's to Vancouver, it's easy to understand that Canada is truly a forest nation. It is the home to 10% of the planet's forest cover (Natural Resources Canada 2005b), 25% of its natural forest, and a world-class forestry industry, which has been the country's primary economic sector ever since the beginning of Confederation. In contrast, almost half of the earth's original forest cover is gone, but Canada has 91% of its native forests intact. Within these 401 million hectares of forest, one-quarter is managed mainly for timber production with less then 0.5% harvested annually (Forest Products Association of Canada 2005).

According to Industry Canada (2002), Canada is a global leader in forest product exports, accounting for 21% of international trade. Few countries are as well poised to benefit from the move to forest-based bioproducts and to help lead the global shift toward a cleaner, bio-based economy. Nevertheless, the ultimate success of forest-based bioproducts requires much more than ample forest resources and a well-developed industry. Significant barriers must be overcome if Canada is to increase the use of forest biomass into areas such as bioenergy, to match the current levels found in Europe (Forest Products Association of Canada 2005a). Even though significant progress is being made, current public and private investments in bioproducts and renewable biomass energy in Canada are still relatively low, and are not sufficiently focused on our true forest biomass potential.

Other pieces of the puzzle need to be in place as well. One of these key pieces is a commitment to research, development and commercialization. Innovation in science, technology and socio-economic analyses are vital to improve our understanding of bioproducts and their potential markets. The industry needs to be supported by: 1) the efficient conversion of forest biomass into commodities such as biofuels, bioplastics, bio-chemicals and other non-timber and non-paper materials; 2) agroforestry systems that allow the co-management of non-timber forest products with timber resources; 3) different forest bioproduct manufacturing strategies that minimize the environmental costs; 4) networking and partnerships with aboriginal and rural forest communities, as well as the food, pharmaceutical, agricultural, "green" construction, energy and chemical sectors; 5) public perception of the benefits and risks of forest bioproducts; and 6) management schemes and policy framework that are consistent with best practices for the sustainability of the resources. To fill these knowledge gaps, Canada urgently needs enhanced research funding, expertly trained researchers and practitioners, as well as organized collaboration amongst governments, industry, academia and non-governmental organizations.

A second important piece of the puzzle is a strategy for navigating around potential barriers to change. For example, consumer social or ethical concerns about biotechnology might negatively impact any attempt to create a viable market for some bioproducts. Similarly, other potential economic, governance and policy roadblocks to the development of bioproducts exist that need to be recognized and openly addressed. The purpose of this concluding chapter is to discuss these various elements on the path forward to a bio-based economy, and to alert the reader to current developments and resources available to further increase understanding of this emerging sector.

1. THE ROLE OF RESEARCH AND DEVELOPMENT IN THE ECONOMY

Research and development drive innovation. Innovation is the process through which new economic and social benefits are extracted from knowledge. A societal commitment to innovation has become arguably the most important engine of economic growth for developed nations over the last century. Between 1948 and 1985, for example, (Table 13-1) techno-logical progress accounted for half to three-quarters of the increased economic performance of the United States, Japan, the United Kingdom, West Germany and France when compared with the relative contributions of labor and capital (Boskin and Lau 1992). Economic growth is supported first by innovation and then by capital.

Table 13-1. The Relative Contribution of Capital, Labor and Technological Progress to Economic Growth.[1]

Country	Capital	Labor	Technical Progress
France	28	−4	76
West Germany	32	−10	78
Japan	40	5	55
United Kingdom	32	−5	73
United States	24	27	49

Source: Boskin and Lau (1992).
[1]Expressed as a percentage of the total growth in gross domestic product for five countries between 1948 and 1985.

As such, it's important to engage in research and development to ensure innovation and sustained economic growth in Canada. In fall 2003, then incoming Prime Minister Paul Martin challenged Canada to create "*an economy driven... by individual ingenuity and creativity.*" He urged Canadians to recognize "*the potential of transformative technologies.*" We must "*create sustainable, long-term growth in the new economy and position Canada as a global high-tech leader,*" he said. The 2004 speech from the throne made a similar case: "*We want a Canada that is a world leader in developing path-breaking technologies of the 21st century - biotechnology, environmental technologies ... [to] ensure that our knowledge investment is translated into commercial success. Get our ideas and innovations out of our minds and into the marketplace*" (Canada Governor General 2004).

This commitment can take two forms. One is financial–as direct investment or as tax incentives–to help industry, governments and univer-sities engage in research and development. The other is investment in education and training to ensure we have enough scientists, technicians and graduates from a variety of disciplines to build effective bioproducts research and development programs, as well as to understand the potential impact of these new technologies and products on Canadian society. Particularly, it is important to educate researchers to bridge the gap between the academic sector and the industrial sector. Indeed, many good ideas remain unimplemented because the transformation of knowledge into implementation depends on entrepreneurs.

2. RESEARCH AND DEVELOPMENT IN CANADA'S FOREST SECTOR

Since its deeply rooted beginnings in Canada's economic, social and cultural history, the forestry sector has relied heavily on technological progress to fuel its growth and maintain its competiveness. Often criticized for lack of spending on research and development, the forest products industry has nonetheless invested billions over the past 15 years in research and development to improve environmental performance, reducing waste and lowering production costs. For instance, industry research and development has helped reduce greenhouse gas (GHG) emissions by 28% since 1990, the largest reduction of any industrial sector in the country. The sector surpassed its Kyoto targets by more than four times during this period, while increasing production by over 30%. The forest products industry is also the first to commit to a further 15% reduction in GHGs, the equivalent to taking 300,000 cars off the road by 2010 (Lazar 2005).

Similar commitment to research has helped Canada's traditional pulp and paper sector embrace the potential of biomass energy (55% of its own energy consumption is from biomass), and it is now Canada's largest industrial source of co-generated power, according to the Forest Products Association of Canada (Forest Products Association of Canada 2005) and the Canadian Bioenergy Association (Canadian Bioenergy Association 2005). Between biomass-fired cogeneration (combined heat and power) and small hydro generation, the sector produces enough renewable energy to power Vancouver, or Calgary and Edmonton combined.

The industry is poised to advance even further as a number of breakthrough technologies currently under development hold the potential to dramatically increase this energy potential, reducing its reliance on GHG-emitting fossil fuels. With the right policy framework, the industry has the potential to be a net exporter of renewable energy (Lazar 2005).

Despite these advances, as a nation, Canada suffers a competitive disadvantage in research intensity. Governments in other countries are investing more extensively in research and development. Canadian R&D is 0.6% of gross industry revenues, compared with 2% in Scandinavian countries and 1.5% in the United States (Government of Canada 2002b). Within the federal government, R&D spending (as a percentage of gross industry revenues) is currently lower in forestry than comparable spending in other resource sectors in Canada. In 1996, for example, the Canadian Forest Service of Natural Resources Canada funded research and development activities to an amount equivalent to 0.16% of forest industry sales. This compares to 1.33% of related industry sales spent on R&D by Agriculture and Agri-Food Canada and 6.83% spent by the Department of Fisheries and Oceans during the same period (Natural Resources Canada 1998).

Spending on forest-related research in Canada grew by 27.2% between 1982 and 1999, from $228.3 million to $290.3 million (in constant 1992 dollars). During this period, industry research funding increased by 32.4% and research and development investment by the provinces grew by 175.1%. Meanwhile, federal government spending fell by 30.4% in the same time frame, from $87 million to $60 million (Watts and Kozak 2000).

Large discrepencies exist between accounting of investments made in research and development by industry. The forest industry calculates it spent 1.7% of its revenue on research and development in 2002. According to more stringent definitions of research and development by Statistics Canada, this number is closer to 0.64% (Canadian Forest Innovation Council 2005). Much of this difference can be ascribed to a more generous definition of research and development by industry, which considers downstream activities, such as mill trial, as "research". To compare with other forested nations, private sector expenditures in science and technology as a % of 1997 forest product shipments was 0.00959 in Canada, 0.05404 in the US and 0.124387 in Sweden (Binkley and Forgacs 1997). Total research and development investment by the forest sector as a whole as defined by Statistics Canada (inclusive of governments) in 2002 was 1.2% of forest revenues (Canadian Forest Innovation Council 2005). It is apparent that, despite past achievements by the industry, funding for research and development in Canada's forestry sector will need to grow if we are to reach the level of commitment shown in many other forestry-intensive nations. It will also need to grow significantly if Canada is to reach its full potential as a world leader in the global bio-based economy.

On the other hand, compared to other developed countries, Canada is generous in its support for corporate research and development through tax subsidies and other tax incentives (Fig. 13-1). For instance, the federal corporate income tax system provides a number of significant tax programs for companies conducting research and development. One of these is the federal Scientific Research and Experimental Development program (SRED), which allows companies to deduct the costs of all qualifying research and development expenditures as well as all spending on research-related equipment and machinery. Further, the program provides an investment tax credit (20% for large companies and 35% for small businesses) on all qualifying SRED expenses incurred in Canada.

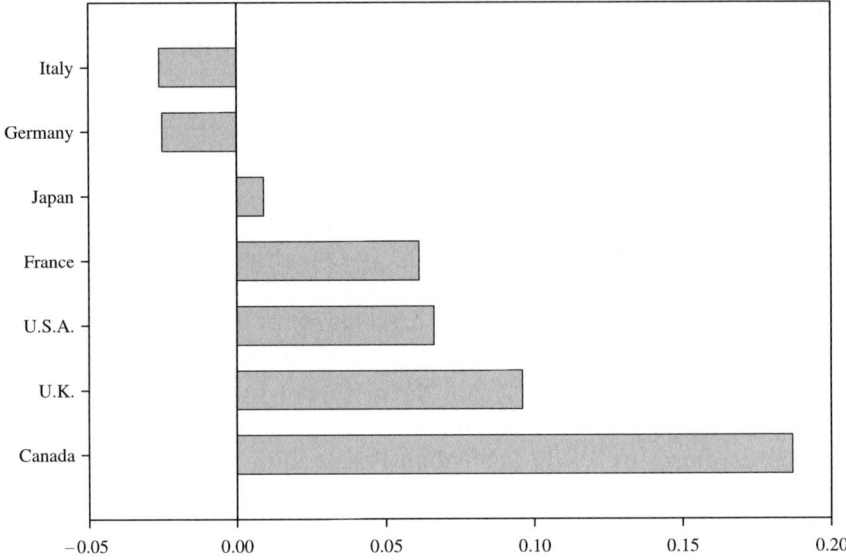

Figure 13-1. Rate of corporate tax subsidies per US$1 of spending on research and development for seven countries.

Adapted from: Organisation for Economic Co-operation and Development (2003).

The provinces generally follow federal rules for allowing deductions for the costs of research and development (Table 13-2). Currently, British Columbia, Saskatchewan, Manitoba, Ontario, Quebec, New Brunswick, Nova Scotia and Newfoundland provide provincial tax breaks for corporate research and development. Alberta and Prince Edward Island do not. Corporations in these two provinces are eligible only for federal tax subsidies.

Tax incentives notwithstanding, research and development spending by the forest industry in Canada remains limited. According to a report prepared by the Forest Coalition for the Advancement of Science and Technology (FORCAST), biotechnology is one of the more impoverished areas of research and development in Canada's forestry sector (Watts and Kozak 2000). Biotechnology research funding for all industries in the natural resource sector was $10.6 million in 1995, up from $4.4 million in 1989.

This is still an order of magnitude less than the $193.1 million spent in 1995 within the health sector (up from $72.7 million in 1989). The FORCAST report concludes that biotechnology – an area of science and technology that is a key tool in the development of forestry bioproducts – receives the lowest financial commitment of any forestry-related research.

Table 13-2. Provincial Comparisons of After-tax Costs (ATC) of $1 Research and Development (R&D) Spending by Corporations as well as B-index Values[1].

Province	ATC of $1 R&D Expenditure		B-index[1] ATC/(1-tax rate)	
	Large Firm	Small Firm	Large Firm	Small Firm
B.C.	0.448	0.473	0.730	0.604
Alta.[2]	0.527	0.547	0.831	0.676
Sask.	0.484	0.462	0.713	0.585
Man.	0.438	0.461	0.719	0.584
Ont.	0.507	0.464	0.787	0.591
Que.	0.482	0.288	0.699	0.369
Nfld.	0.517	0.477	0.709	0.582
P.E.I.[2]	0.581	0.537	0.825	0.676
N.B.	0.461	0.497	0.757	0.614
N.S.	0.444	0.477	0.717	0.582

Source: Statistics Canada (1999).

[1] B-index = ATC/(1-tax rate).

[2] Province not offering R&D tax credit.

Indeed, Industry Canada and BioProductsCanada recommend investment over the short and long term in biotechnology research, skills training and bioproduct innovation (Industry Canada 2004). Improvements in marketing and public outreach, sharing of data and best practices amongst suppliers, institutional collaboration, and new policy mechanisms were also suggested.

3. THE ROLE OF EDUCATION AND TRAINING

The Canadian forest industry relies heavily on skilled, technically trained researchers. On a national scale Canada does not suffer from a lack of scientific or technically skilled people to fill the current demands of Canadian industry (Advisory Council on Science and Technology 1999). Indeed, Canadian universities now supply more graduates than the workforce generally requires (Human Resources Development Canada 1998). On a regional scale, however, there are shortages of skilled personnel.

Moreover, in fast growing sectors – including the biotechnology industries that are important for the development of forest bioproducts – recruiting people to fill areas of specific technical expertise has frequently been difficult (Advisory Council on Science and Technology 1999). Biotechnology companies lament that finding people who combine scientific and technical expertise with management ability is especially problematic.

Indeed, Canadian students are usually not cross-trained in the diverse disciplines demanded by the bio-based economy, the converging fields of biology, chemistry and engineering, nor do students gain adequate exposure to the socioeconomic implications of their work. To meet these challenges, multi-sector stakeholders need to work together in partnership to forge opportunities that transform and expand the user sector, increase consumer confidence, and retain highly skilled workers in Canada.

Post-secondary education will be a requirement for the vast majority of new jobs in the 21st century, according to Canada's Innovation Strategy (Government of Canada 2002b). Universities and colleges are therefore expected to play an increasingly important role in supplying a scientifically literate and technically skilled workforce for a bio-based economy. Investment in post-secondary education is therefore likely to become an important component of Canada's economic performance in the years ahead.

Meanwhile, Canada's funding commitment to post-secondary education has diminished by 30% (in relative terms) over the past two decades, widening the funding gap for higher education between the country and its largest trading partner, the US, where funding has increased during the same period by 20% (Association of Universities and Colleges of Canada 1999). Further, the overall budget for the Natural Science and Engineering Research Council (NSERC)–Canada's premier granting agency for university research programs in non-health-related science and technology– has also declined since 1993, while the university research investments of other G7 nations have increased (Association of Universities and Colleges of Canada 1999). Canada will need to reverse these trends if it hopes to support university-based research and development to a level that is competitive on the world stage.

A coherent national policy on university intellectual property is also expected to help ensure support for university research programs in the research-intensive economy of the future. Today, Canada's federal granting councils–the Natural Science and Engineering Research Council (NSERC), the Social Sciences and Humanities Research Council (SSHRC), and the Canadian Institutes of Health Research (CIHR)–invest more than $940 million in university research annually. Yet, they only require researchers to disclose publications and patents generated from grants, and the councils claim no ownership of any intellectual property. The result has been a wide diversity of practices and complexity of approaches governing IP ownership and disclosure at Canadian universities. According to a report by the federal government's Expert Panel on the Commercialization of University Research

(Expert Panel on the Commercialization of University Research 1999), this absence of a coherent national policy is resulting in the loss of commercialization opportunities, intellectual property benefits lost to other countries, costly litigation, and limitations for the longer-term innovative potential of small, knowledge-intensive Canadian firms.

4. IDENTIFYING POTENTIAL BARRIERS TO FOREST BIOPRODUCTS

Forest bioproducts promise many novel non-timber products as well as clean, renewable sources of energy, chemicals and materials. They offer the opportunity for Canada to make the most of its abundant forest resources, including organic material that may currently be underused or wasted. As hopeful as these promises are, they are just as generously accompanied by concerns. Questions abound regarding the environmental, social, economic and ethical implications of bioproducts that need to be acknowledged. If the forest bioproducts industry wants to ensure the success of promising, non-traditional markets, it must also provide transparent and balanced information to risk-wary consumers.

Humankind's record at managing natural resources is grounds for concerns about the most important outstanding environmental issues associated with the development of the forest bioproducts industry. For centuries, human societies have used biomass resources – and forest resources, in particular – to heat their homes, to cook their food, and to make tools and shelter. As a result, forests in countries throughout the world have been devastated. Around half of the earth's original forest cover has been removed by anthropogenic disturbance. Large portions of valuable habitats have been extirpated. Air pollution from wood or charcoal smoke has been an ongoing problem in both urban and rural areas ever since the Middle Ages. This legacy needs to be taken into account as we plan the expansion of the bioeconomy.

Current environmental concerns include the impact of forest bioproducts on pollution, biodiversity or water resources. For instance, the increased use of pesticides or fertilizers that might be required for growing bioenergy feedstocks are a potential source of chemical and microbial contamination of groundwater, may contribute to eutrophication of surface waters, or pollute the soil and air. Monocultures dedicated toward biomass might also threaten habitat quality for many species of wildlife with ecological requirements for old growth habitats, while it might favor species that can tolerate and/or thrive in such ecosystems. Intensive tree farming, while perhaps stimulating rural economies, could also place high demands on area and community water supplies.

A large-scale industrial harvest of trees which outstrips our ability to regenerate forests or which imperils forested wild lands will no doubt *not* be tolerated by an increasingly informed public. The industrial use of forest and other biological resources thus will need to be clean and sustainable if it is going to be a viable alternative to the use of petroleum.

Another environmental question for forest bioproducts concerns their impact on climate change. Forest bioproducts are often cited as a means to help fight this global problem by providing an alternative to the fossil fuels responsible for much of our greenhouse gas emissions. On the other hand, the amount of airborne carbon dioxide – the most abundant greenhouse gas – could actually increase if too many trees are used or if ecosystems (including old growth, forest floor litter and soil) that store carbon in large quantities are sacrificed to grow trees with lower carbon densities.

The relationship between forest bioproducts and climate change becomes more complicated when one considers how increased numbers of fires and insect infestations associated with a warmer climate may reduce biomass supply. While the signs of impacts of climate change on Canadian forests are already visible, the opportunities to productively adapt to these impacts are less visible and less understood. The higher risk of forest fire and insect damage will require as yet unknown strategies and practices by the forest industry and provincial/territorial forest management agencies to avoid declines in timber quality and quantity (Lazar 2005).

A 2004 *Primer on Bioproducts* published by Pollution Probe and BIOCAP Canada Foundation (2004) addresses many of these complex issues. In particular, the primer examines environmental, economic, social and ethical issues identified by representative stakeholders as being of concern to Canadian citizens. The most problematic areas of concern focus on biotechnology, especially as applied to our forests and natural environment.

For example, some forest bioproduct industries may rely on genetic engineering to isolate and enhance the performance of microorganisms and their catalyzing enzymes, or to produce fast-growing or pest-resistant trees. The transfer of DNA between tree species can be used to increase tree production capacity or to enhance performance as biomass raw material for industry. Public acceptance of these sometimes-controversial techniques will likely depend on the perceived benefit of specific bioproducts for individual consumers and the well being of society as a whole.

Other ethical considerations facing environmental biotechnology research are discussed in a yet another *Primer for Scientists*, published in 2001 by the Ottawa-based Institute on Governance (Institute On Governance

2001). Some ethical and legal questions arising from the development of bioproducts could include balancing precaution against inaction in the development of technologies (affecting innovation), establishing a system to determine whether or not the means of research and development for bioproducts is morally justifiable (with respect to humans, animals, and eco-systems), and assessing the potentially conflicting interests and obligations of bioproducts researchers and their funding sources. These value-laden decisions will influence the rules, principles and ways of thinking to help determine whether Canadians are on the right track – morally and philoso-phically justified – in the development of a bioproducts industry and regu-latory framework.

Organizations involved in the ethical consideration of research and development that could have an impact on forest bioproduct development include the Canadian Biotechnology Advisory Committee (CBAC), an agency that provides advice to the federal government on ethical, social and regulatory aspects of biotechnology, and the International Bioethics Committee of UNESCO. Yet more economic and policy–related questions raised by forest bioproducts include confusion around international trade. Current trade agreements affecting Canada, such as the North American Free Trade Agreement (NAFTA) and those of the World Trade Organization (WTO) may not be adequate to consider forest bioproducts. International agreements on safety, monitoring, processing, regulating intellectual property and government financing will need to be updated.

Ultimately, the benefits and the concerns raised by forest bioproducts and biotechnology make it critical that Canadians are well informed and able to participate in discussions about future directions and stewardship of their national forest legacy. This book, as well as the many resources cited within, aim to inform and facilitate such needed debate.

5.   THE FUTURE

"Where You Find BioProducts You Find the Future" is the caption on the first annual industry guide published in February 2004 entitled, *"Canadian BioProducts From Renewable Resources 2004"* (Contact Canada 2005). In addition, Contact Canada's search engine directory catalogues Canadian bioproduct companies, along with many other supporting organizations, research institutes and government agencies. The first wave of Canadian bioproduct companies includes many that profit from residual forestry biomass.

Of course, the "traditional" forest products industry already plays a significant role in the Canadian economy. It contributed $33.7 billion to

Canada's gross domestic product in 2003 (Natural Resources Canada 2005b). It employs more people in more parts of the country than any other industrial employer. About 3,550 forestry-related operations across the country directly employ 376,300 Canadians and provide indirect employment for another 700,000. About 350 Canadian communities depend on the industry. It is important that the emerging bioproducts industry complements this "traditional" industry and does not compete with it.

Non-timber forest bioproducts offer the promise of making Canada's forestry sector even more valuable as a vital source of clean, renewable alternatives to fossil fuels in the industrial production of energy, chemicals and materials, along with the many other value-added foods, pharmaceuticals and specialty products described in this book. They capitalize on markets and industrial infrastructures that are already in place in Canada. That means demand almost certainly exists. As long as these bioproducts are priced competitively, the forest sector is in a remarkable position to take advantage of new consumer interests, as well as growing concerns about the impact of fossil fuel use on our energy security and on environmental integrity.

Yet, how these bioproducts will eventually affect the environment or the lives of Canadians has yet to be determined. There is a need to compare and analyze the impacts of different forest bioproduct technologies since there is little peer-reviewed material dealing with the forestry sector. The case for more extensive research has been made, and questions concerning the collection of biomass raw materials, the distribution of bioproducts and the ultimate socio-economic impact still need to be answered.

National multi-sectoral, not-for-profit, development organizations such as BioProducts Canada Inc. (BPC) and its regional and provincial counterparts are helping to raise the public awareness and profile of emerging bioproduct industries from all natural resource sectors. As was made clear by the joint BPC and Industry Canada *Innovation Roadmap*, bioproducts are a welcome catalyst in support of each of the three pillars of social, economic and environmental sustainability.

The roadmap concludes, "*Governments need to create a national demonstrations facility to establish and network key technologies. The biotechnology industry needs to be marketed or branded better. Our message needs to be widely communicated. A strong public engagement program needs to accompany the action plan from this roadmap.*" (Industry Canada 2004).

To move forward, forest science, policy and practices must collectively address future uncertainties, including the impacts of climate change, on the forest and forest sector, in order to build adaptive strategies that:

- Ensure the continued flow of environmental, economic and social benefits from our forests (*i.e.*, sustainable forest management).
- Build resiliency to environmental, economic and social risks into forest management and community decision-making.
- Recognize the threats and opportunities that climate change poses to the fulfillment of Canada's international obligations in the forest sector, *e.g.*, those pertaining to biodiversity and the Kyoto Protocol.

In the final analysis, forests have many more values than the production of wood and the creation of jobs. Issues such as clean air, pure water, ecosystem and community health, along with the spiritual, cultural and recreational values that Canadians hold for their forests, may all be placed in jeopardy without careful planning. With innovative partnerships working together to capture Canada's natural advantage, Canadians can hope to achieve the vision of a sustainable bio-based economy for the world's leading forest nation.

References

Adam, K., Kuepper, G.L. and Diver, S. 2000. *Organic blueberry production*. Appro-priateTechnology Transfer for Rural Areas (ATTRA). Retrieved Feb. 2004 from: http://attra.ncat.org/attra-pub/PDF/blueberry.pdf.

Advisory Council on Science and Technology 1999. *Stepping up. Skills and opportunities in the knowledge economy*. Retrieved Sept. 2005 from: http://acst-ccst.gc.ca/skills /finalrep_html.

Agriculture and Agri-Food Canada 2000. *The health of our water. Toward sustainable agriculture in Canada*. Coote, D.R. and Gregorich L.J. (eds). Publication 2020/E. Cat. No. A15-2020/2000E. ISBN 0-662-28489-5. (Online Version: http://res2.agr.gc.ca/publications/ hw/PDF/eifor.pdf).

Agriculture and Agri-Food Canada 2000a. *The Canadian ginseng industry: Preparing for the 21st century*. (Online Version: http://atn-riae.agr.ca/can/e2765.htm).

Agriculture and Agri-Food Canada 2000b. *Profile of the Canadian cranberry industry*. Market & Industry Services Branch. Retrieved Oct. 2004 from: http://www.agr.ca/.

Agriculture and Agri-Food Canada 2001. 2000/2001 *Canadian maple products situation and trends*. Market and Industry Services Branch, Horticulture and Special Crops Division. Ottawa, ON. (Online Version: http://www.agr.gc.ca/misb/hort/pdf/maple_eng.pdf).

Agriculture and Agri-Food Canada 2002. *Non-food/non-feed industrial uses for agricultural products, phase 1*. PRA Inc. Information Into Strategy and CANUC (Canadian Agricultural New Uses Council), Prepared for Agriculture and Agri-Food Canada, June 19, 2000. 85 pp.

Agriculture and Agri-Food Canada 2002a. *Functional foods and nutraceuticals: Market, industry and distribution*. Retrieved Jan. 2003 from: http://www.agr.gc.ca/food/nff/ fnnmarket/ffnmrket.html.

Agriculture and Agri-Food Canada 2002b. *Canadian floriculture nursery and Christmas tree situation and trends*. Market Industry & Services Branch. Retrieved Dec. 2004 from: www.agr.ca.

Agriculture and Agri-Food Canada 2003. *2002/2003 Canadian honey situation and trends*. Market and Industry Services. Horticulture and Special Crops Division. Ottawa, ON. (Online Version: http://www.agr.gc.ca/misb/hort/trends-tendances/pdf/honey02-03_e.pdf).

Agriculture and Agri-Food Canada 2003a. *Canada's organic industry.* Canada's Agriculture, Food and Beverage Industry. Agri-Food Trade Service, Products and Suppliers Fact Sheets. Retrieved Mar. 2005 from: http://ats-sea.agr.gc.ca/supply/3313_e.htm.

Agriculture and Agri-Food Canada 2003b. *Organic agriculture standards development in Canada.* (Online Version: http://ats-sea.agr.gc.ca/can/e3468.htm).

Agriculture and Agri-Food Canada 2003c 2002/2003 *Canadian maple products - situation and trends.* Market and Industry Services Branch, Horticulture and Special Crops Division.Ottawa, ON. (Online Version: http://www.agr.gc.ca/misb/hort/trends-tendances/pdf/map02_03_e.pdf).

Agriculture and Agri-Food Canada 2003d. *2002/2003 Canadian fruit situation and trends including apples, tender fruits, grapes (Vinifera) and berries.* Market and Industry Services Branch, Ottawa, Ontario. (Online Version: http://www.agr.gc.ca/misb/hort/sit/pdf/fru_02_03_e.pdf).

Agriculture and Agri-Food Canada 2003e. *Developments in berry production and use.* Winnipeg, Manitoba. ISSN 1207-621X. AAFC No. 2081/E. (Online Version: http://www.agr.gc.ca/mad-dam/e/bulletine/v16e/v16n21_e.htm).

Agriculture and Agri-Food Canada 2003f. *Special crops. Ginseng statistics.* (Online Version: http://www.agr.gc.ca/misb/spcrops/sc-cs_e.php?section=stats&page=ginseng).

Agriculture and Agri-Food Canada 2003g. *Canada's ginseng industry.* Markets and Trade Special Crops Information. Retrieved March 2004 from: http://www.agr.gc.ca/misb/spcrops/sc-cs_e.php?page=ginseng.

Agriculture and Agri-Food Canada 2003h. *Canada's wild rice industry.* Special Crops. (Online Version: http://www.agr.gc.ca/misb/spcrops/sc-cs_e.php?page=wildrice-rizsauvage).

Agriculture and Agri-Food Canada 2003i. *Canada's functional foods and nutraceuticals industry.* Retrieved March 2004 from: http://ats-sea.agr.gc.ca/supply/3312_e.pdf.

Agriculture and Agri-Food Canada 2003j. *Pesticide use in Canada.* Retrieved Aug. 2005 from: http://www.agr.gc.ca/index_e.phtml.

Agriculture and Agri-Food Canada 2004. *Prairie farm rehabilitation administration.* Retrieved Jan. 2005 from: http://www.agr.gc.ca/pfra/main_e.htm.

Agriculture and Agri-Food Canada 2004a. *Canada's maple syrup industry.* Market and Industry Services Branch. (Online Version: http://ats-sea.agr.gc.ca/supply/3310_e.htm).

Alberta Agriculture, Food and Rural Development 2000. *The market for herbs and essential oils.* Retrieved Mar. 2003 from: http://www.agric.gov.ab.ca/economic/market/herbsarticle. html.

Albertson, D. M. and Pope, K. M. 1999. *Paper sludge - Waste disposal problem or energy opportunity.* Energy Products of Idaho. Retrieved Apr. 2004 from: http://www.energy-products.com/Documents/SLUDGPA4a.PDF.

Alexander, S.J. 2001. *Who, what, and why: The products, their use, and issues about management of non-timber forest products in the United States.* In: Forest Communities in the Third Millennium: Linking research, business and policy toward a sustainable non-timber forest product sector. (Online Version: http://www.ncrs.fs.fed.us/pubs/gtr/gtr_nc217.pdf).

Alexander, S.J., Weigland, J. and Blatner, K.A. 2002. "Nontimber forest product commerce." In: *Nontimber Forest Products in the United States.* Jones, E.T., McLain, R.J. and Weigland, J. (eds.). University Press of Kansas, Lawrence, K.S. pp. 115-150.

Alternative Medicine Review 2000. *Monograph: Larch arabinogalactan.* Alternative Medicine Review5(5): 2000. Retrieved Jan. 2003 from: http://www.thorne.con/pdf/journal/5-5/larch-monograph.pdf.

Alternative Medicine Review 2001. *Monograph: Vaccinium myrtillus (bilberry).* Alternative Medicine Review 6(5): 2001. Retrieved Jan. 2003 from: http://www.thorne.com/altmedrev/bilbery6-5.html.

Alternative Medicine Review 2001a. *Monograph: Plant sterols and sterolins.* Alternative Medicine Review 6(2): 2001. Retrieved Jan. 2003 from: http://www.thorne.com/altmedrev/plant6-2.html.

American Forests 2000. *Pass the birch syrup please.* American Forests Magazine. Wntr. 2000. Retrieved Sept. 2004 from: http://www.findarticles.com/p/articles/mi_m1016/is_4_105/ai_58381494.

Amey, A. 2003. *Can Canada fast-track on carbon sinks?* C3 Views 3: 3-4.

Anon. 2002. *Beneath the trees.* The non-timber forest products demonstration project newsletter. Issue 5. Royal Roads University.

Arbo 2004. *Lignosulfonates: A complete range of lignin products.* Retrieved June 2005 from: http://www.arbo.ca/pages/en/aboutus.html.

Arborvitae Environmental Services 1997. *Commercialization of special forest products in the lake Abitibi model forest and region.* Final project report: Assessment of commercialization potential of wild mushrooms, aromatic oils, and decorative twigs and branches. Toronto, ON, 60 pp.

Archambault, E. 2004. *Towards a Canadian R&D strategy for bioproducts and bioprocesses.* Science-Metrix Canadian R&D Biostrategy. Prepared for National Research Council of Canada. Retrieved June 2004 from: http://www.science-metrix.com/.

Arif, B. 1995. "The nature of insect baculoviruses." In: *Research in Canada. Recent Progress in Forest Biotechnology in Canada.* PI-X-120. pp. 108-113.

Ashbury Biologicals Inc. 2003. *Products: tanacet 125® for migraine prevention.* Retrieved Jan. 2003 from: http://www.ashburybio.com/products/htm.

Association for Temperate Agroforestry 2004. *An introduction to temperate agroforestry.* University of Missouri, Columbia, MO 65211. Retrieved Jan. 2005 from: http://www.aftaweb.org/.

Association for Temperate Agroforestry 2004a. *Alley cropping with hybrid poplar may profit UK farmers.* University of Missouri, Columbia, MO 65211. Retrieved Jan. 2005 from: http://www.aftaweb.org/entserv1.php?page=29.

Association of Equipment Manufacturers 2001. *New diesel fuels: They are in your future for nonroad equipment.* Retrieved Feb. 2004 from: http://www.aem.org/Technical/PDF/diesel-fuels-rpt.pdf.

Association of Universities and Colleges of Canada 1999. *Trends: The Canadian university in profile.* Ottawa, Canada.

Australian Government. 2001. *R&D plan for essential oils and plant extracts 2002-2006.* Rural Industries Research and Development Corporation. Retrieved Nov. 2004 from: http://www.ridc.gov.au/pub/essentoi.html.

Automotive News 2002. *Nissan speeds fuel cell program.* 12/23/2002 Automotive News. Crain Communications Inc.

Avorn J., Monane M, Gurwitz J.H., Glynn R.J., Choodnovskiy I. and Lipsitz L.A. 1994. *Reduction of bacteriuria and pyuria after ingestion of cranberry juice.* The Journal of the American Medical Ass. Vol. 271, No. 10, Mar. 9, 1994. (Online Version: http://jama.ama-assn.org/cgi/content/abstract/271/10/751?maxtoshow=&HITS=10&hits=10&RESULTFORMAT=&fulltext=Cranberry+Juice&searchid=1122296750719_977&stored_search=&FIRSTINDEX=0&journalcode=jama.

Battelle 2005. *Mushrooms: Higher macrofungi to clean up the environment.* Battelle environmental updates. Retrieved Aug. 2005 from: http://www.battelle.org/Environment/publications/ EnvUpdates/Fall00/article4.html.

Beardmore, T. 1995. "The molecular biology of tree seed development." In: *Research in Canada. Recent Progress in Forest Biotechnology in Canada.* PI-X-120. pp. 307-319.

Beckwith, A.F., Roebblelen, P. and Smith, V.G. 1983. *Red pine plantation growth and yield tables.* Ont. Min. Natur. Resources, Forest Res. Br., Forest Res. Rep. No. 108. 70 pp.

Belsito, K. and Winterhalder, K. 2000. *Composted papermill sludge as a dry cover and vegetative growth substrate for acid-generating tailings and acidic, metal-contaminated soils.* In: Etmanski, A. (ed.), Proc. 25th Ann. Meeting, Canadian Land Reclamation Association and 4th Meeting of the International Affiliation of Land Reclamationists, Edmonton, Alberta, September 16-21 2000.

Biby, G. 2005. *Degradable polymers.* Retrieved June 2005 from: http://www.icma.com/info/polymers.htm.

Bidleman, R. 2000. *Vaccinium myrtillus, bilberry, huckleberry, whortleberry, etc.* Retrieved July 2005 from: http://herb.com/files/bilberry.html.

Binkley, C.S. and Forgacs, O.L. 1997 *Status of forest sector research and development in Canada.* Report commissioned by the Canadian Forest Service.

Bio Development International 2003. *Le calorie S.p.a.* Retrieved May 2004 from: http://www.bdi-us.com/crnt_lecal.html.

BIOBUS Project 2003. *Biodiesel: Demonstration and assessment with the Societe de transport de Montreal (STM).* Final report. Retrieved Feb. 2005 from: http://www.stcum.qc.ca/English/info/a-bus-final.pdf.

BIOCAP Canada Foundation 2004. *Brief issue 8, September 2004.* Retrieved Aug. 2005 from: http://www.biocap.ca/files/Briefs/A_consequence.pdf.

BIOCAP Canada Foundation 2004a. *Understanding varying year-to-year greenhouse gas increases: critical to making reliable estimates of Canada's carbon balance.* BIOCAP Brief, Issue 1. (Online Version: http://www.biocap.ca/files/Briefs/Inter_var.pdf).

BIOCAP Canada Foundation 2005. *Carbon levels continue to rise.* In the News. April 6, 2005. (Online Version: http://www.biocap.ca/files/In The News/Volume 1/Issue4.pdf).

BIOCAP Canada Foundation 2005a. *Carbon dioxide price continues to surge.* In the News. June 22, 2005. (Online Versi on: http://www.biocap.ca/files/In The News/ VolumeI/Issue5. pdf).

Biocontrol Network 2002. *Research programs.* Retrieved Aug. 2005 from: http://www.biocontrol.ca/ english/start_s.html.

Biswas, S. and Vashishtha, N. 1998. *Xylitol: Technology & business opportunities.* Chemical Engineering World, 33(1):103-108 (Online Version: http://www.tifac.org.in/news/view6.html).

Bombay, H.M. 2001. *Introductory remarks from the national Aboriginal forestry association.* In: Forest Communities in the Third Millennium: Linking Research, Business, and Policy Toward a Sustainable Non-Timber Forest Product Sector. (Online Version: http://www.ncrs.fs.fed.us/pubs/gtr/other/gtr-nc217/).

Booth, D. 2005. *Perspectives on article 3.4.* Power Point Presentation to the BIOCAP Canada Conference, Feb. 3, 2005. (Online Version: http://www.biocap.ca/images/pdfs/conferenceSpeakers/Booth_D.pdf).

Bord na Móna 2001. *Peat for domestic heating.* Bord na Móna Energy Limited. Leabeg, Tullamore, Co. Offaly, Ireland. Retrieved Apr. 2004 from: http://www.bnm.ie/downloads/peat_for_domestic_heating.pdf.

Boskin, M. J. and Lau, L.J. 1992. "Capital, technology, and economic growth." In: *Technology and the Wealth of Nations.* Stanford University Press, Stanford, US. pp. 17-55.

Boxall, P.C., Murray, G. and Unterschultz, J. 2002. *The potential for non-timber products from the Boreal forest: An exploration of Aboriginal opportunities.* In: Proceedings of the

Sustainable Forest Management Network Conference, "Advances in Forest Management: From Knowledge to Practice." T. Veeman, P. Duinker, B. Macnab, A. Coyne, K. Veeman, G. Binstead and D. Korber (Eds). Edmonton, Alberta. November 13-15, 2002. pp. 234-239.

Brakenhielm, S. and Liu, Q. 1998. *Long-term effects of clear-felling on vegetation dynamics and species diversity in a boreal pine forest.* Biodiversity and Conservation 7:207-220.

British Columbia Ministry of Agriculture, Food and Fisheries 2004. *BC berry and nut production ranked by value - 2002.* British Columbia Ministry of Agriculture, Food and Fisheries. Agriculture Statistics. Retrieved Mar. 2005 from: http://www.agf.gov.bc.ca/stats/berries/49-02.htm.

Brubacher, D. 1999. *Non-timber forest products: exploring opportunities for Aboriginal communities.* Prepared for Mitigaawaaki Forestry Co-operative by D. Brubacher & Associates.

Calvin, M. 1985. *Fuel oils from higher plants.* Department of Chemistry and Lawrence Berkeley Laboratory. University of California. Berkeley, CA.

Cameron M. 2001. *Establishing an Alaskan birch syrup industry: birch syrup-it's the Un-maple!*[TM]. In: Forest Communities in the Third Millennium: Linking research, business and policy toward a sustainable non-timber forest product sector. (Online Version: http://www.ncrs.fs.fed.us/pubs/gtr/other/gtr-nc217/).

Campbell, A.G. and Tripepi, R.R. 1992. *Logyard residues: products, markets, and research needs.* Forest Prod. J. 42(9):60-64.

Canada Governor General 2004. *Speech from the Throne, Feb. 2, 2004.* ISBN 0-662-67926-1. Cat. No. SO1-1/2004. Retrieved Sept. 2005 from: http://pm.gc.ca/grfx/docs/sft_fe2004_e.pdf.

Canadian Bioenergy Association 2005. *Bioenergy.* Retrieved Sept. 2005 from: http://www.canbio.ca.

Canadian Council of Forest Ministers 1997. *Criteria and indicators of sustainable forest management in Canada: Technical report 1997.* Cat. Fo75-3/6-1997E. (Online Version: http://www.ccfm.org/ci/pdf/tech/ci_e.pdf).

Canadian Council of Forest Ministers 2000. *Criteria and indicators of sustainable forest management in Canada: National status 2000.* Cat. F075-3/6-2000E. (Online Version: http://www.ccfm.org/ci/pdf/ns2k/full_report_e.pdf).

Canadian Council of Forest Ministers 2001. *Forest 2020, a dialogue with Canadians.* (Online Version: http://www.ccfm.org/forest2020/index.html).

Canadian Environmental Law Association 2000. Case Study #2: *Regulating pesticides to protect children's health (environmental standard setting and children's health).* Retrieved Aug. 2005 from: http://62.44.8.131/publications/cardfile.shtml?x=1110.

Canadian Environmental Protection Act 1999. *Laws. Consolidated statutes and regulations.* Retrieved Aug. 2005 from: http://laws.justice.gc.ca/en/c-15.31/29610.html.

Canadian Food Inspection Agency 2005. *About the CFIA.* Retrieved Aug. 2005 from: http://www.inspection.gc.ca/english/toce.shtml.

Canadian Forest Innovation Council 2005. *Research expenditures in the Canadian forest sector.* Unpublished. Report prepared by J. Williams, F.C. Pollet, J. Rogers, P. Griss. 22pp.

Canadian Forest Service 2002. *Managing your woodland. A non-forester's guide to small-scale forestry in British Columbia.* ISBN 0-7726-4776-3.

Canadian Forest Service 2004. *Pathology.* Natural Resources Canada. Retrieved Aug. 2005 from: http://www.nrcan-rncan.gc.ca/cfs-scf/science/resrch/pathology_e.html.

Canadian Forest Service 2005. *Forest 2020 Plantation demonstration and assessment initiative.* Retrieved Sept. 2005 from: http://www.nrcan.gc.ca/cfs-scf.

Canadian Forest Service 2005a. *Forest carbon accounting.* Retrieved Aug. 2005 from: http://carbon.cfs.nrcan.gc.ca.

Canadian General Standards Board 1999. *Organic agriculture.* National Standard of Canada. CAN/CGSB-32.310-99. Ottawa, Canada. (Online Version: http://www.pwgsc.gc.ca/cgsb/032_310/32.310epat.pdf).

Canadian Heritage 2004. *Section II: Departmental overview.* Retrieved Oct. 2004 from: http://www.canadianheritage.gc.ca/pc-ch/mindep/perf/98-99/chpr/3_e.cfm.

Canadian Honey Council 2001. *Proplis collection: a value added potential.* Hivelights Magazine. Volume 14, number 1. Feb. 2001. (Online Version: www.honeycouncil.ca/users/folder.asp?FolderID=1727).

Canadian Renewable Fuels Association 2004. *Ethanol and biodiesel. Growing cleaner, affordable fuels.* Retrieved Apr. 2004 from: http://www.greenfuels.org.

Canadian Wildlife Service 2005. *Hinterland who's who: benefits of wildlife.* Retrieved Feb. 2005 from: http://www.hww.ca/hww2asp?cid=4&id=221.

CANMET Energy Technology Centre 2003. *Ethanol the "green gasoline".* Natural Resources Canada. Retrieved May 2005 from: http://nrcan.gc.ca/es/etb/cetc/pdfs/ethanol_the_green_gasoline_e.pdf.

Canwest News Service 2004. *Cranberry juice joins wine as a heart-friendly drink.* Sault Star. Oct. 26, 2004.

Carbon Market Analyst Online 2002. *The carbon market trader. Retrieved* July 2005 from: www.pointcarbon.com

Cargill Dow 2004. *Life cycle of NatureWorks PLA^{TM}.* Retrieved July 2004 from: http://www.cargilldow.com/corporate/life_cycle/life_cycle_faqs.asp

Catling, P.M. and Small, E. 2001. *North American wild rice (Zizania species) - a wild epicurean crop.* Biodiversity V2 (3). 2001.

Chapeskie 1990. Indigenous law, state law and the management of natural resources: wild rice and the Wabigoon Lake Ojibway Nation. In: *Law and Anthropology (International yearbook for legal anthropology)* V5. pp. 129 -166.

Charest, P.J. and Klimaszewska, K. 1995. "Tree germplasm preservation using biotechnology." In: *Research in Canada. Recent Progress in Forest Biotechnology in Canada.* PI-X-120. pp. 10-15.

Chilton, J. 1993. *The health benefits of mushrooms.* NAMMEX, North American Medicinal Mushroom Extracts. Retrieved Oct. 2004 from: http://www.nammex.com?MushroomArticles/healthBenefitMushrooms.html.

Chun, Y.W., Klopfenstein, N.B., McNabb, H.S., Hall, R.B. 1988. *Biotechnological application in Populus species.* J. Kor. For. Soc. 77: 467-783

Ciesla, W.M. 1998. *Non-wood forest products from conifers.* Non-wood forest products Technical Papers. Food and Agricultural Organization of the United Nations, Rome.

Clean Fuels Development Coalition 2002. *Methanol.* Retrieved Mar. 2004 from: http://www.ethanol-gec.org/clean/index.htm.

Clean Fuels Development Coalition 2003. *Methanol.* Retrieved March 2003 from: http://www.ethanol-gec.org/clean/cf05.htm.

CleanAir Canada 2003. *Success stories.* February 15, 2003. 41 pp.

Climate Change Central 2002. *Al-Pac ahead of the game.* C3 Views Newsletter 3, 8-10. Retrieved Mar. 2003 from: www.climatechangecentral.com/info_centre/C3Views/c3views_apr02.html.

Conference Board of Canada 2001. *The future cost of health care in Canada, 2000 to 2020: Balancing affordability and sustainability.* 333-01 Detailed Findings by Brimacombe,

G.G., Antunes, P. and McIntyre, J. Retrieved June 2005 from: http://www. conference board.ca/press/documents/FutureHealth.pdf.

Contact Canada 2005. *Canadian bioProducts from renewable resources 2004.* Contact Canada, Ottawa, Canada. Retrieved Sept. 2005 from: http://contactcanada.com/ bioproducts/.

Couzinier, J.P. and Mamatas, S. 1986. Basic and applied research in the pharmaceutical industry into natural substances. In: *Advances in Medicinal Phytochemistry.* Barton, D., Ossis, W.D. (eds.). John Libbey, London, UK, pp. 57-61.

Coyes & Associates International 2003. *Métis settlements general council economic viability strategy, supplementary report "B," peat moss-additional information.* Retrieved Feb. 2003 from: http://www.msgc.ca/.

Cragg, G.M. 1998. *Paclitaxel (Taxol®): A success story with valuable lessons for natural product drug discovery and development.* Med. Res. Rev. 18: 315-331.

Crawford, C. 2001. *Discussion framework: developing biobased industries in Canada.* Prepared for Horticulture and Special Crops Division (HSCD) and Market and Industry Services Branch (MISB). Retrieved Aug. 2004 from: http://www.bioalberta.com/ ims/client/upload/developing%20biobased%20industries%20in%20canada.doc.

Culhane, C. 1995. *Nutraceuticals/functional foods: an exploratory survey on Canada's potential.* Agriculture and Agri-Food Canada. Retrieved July 2005 from: http://www. agr.gc.ca/misb/fb-ba/nutra/nutra/pdf/rpt_e.pdf.

Cunningham, A.B. 1995. *"An Africa-wide overview of medicinal plant harvesting, conservation and health care."* In: Non-Wood Forest Products 11, Medicinal Plants for Conservation and Health Care. Food and Agriculture Organization of the United Nations. 15 pp. (Online Version: http://www.fao.org/docrep/W7261e/W7261e14.HTM).

Czernik, S.R., French, R., Feik, C. and Chornet, E. 1999. *Hydrogen from biomass via fast pyrolysis/catalytic steam reforming process.* Proceedings of the 1999 US DOE Hydrogen Program review. NREL/CP-570-26938. Retrieved May 2004 from: http://www. eere.energy.gov/hydrogenand fuelcells/pdfs/26938j.pdf.

Daigle, J.-Y. and Gautreau-Daigle, H. 2001. *Canadian peat harvesting and the environment, second edition, issues paper, No. 2001-1.* North American Wetlands Conservation Committee. Published in Partnership with Canadian Sphagnum Peat Moss Association and Canadian Wildlife Service, Environment Canada. (Online Version: http://www. peatmoss.com/Issuepap2.pdf).

Dale, O., Gjolsjo, S., Gronlien, H. and Kjostelsen, L. 1998. *Rundballepressing av hogstavfall I skogen for uttak til bioenergy - en pilotstudie. rapport fra skogforskningen.* Norsk institutt for skogforshning. Hogskolevn, Norway. 12, 1432 As.

D'Amico, V. 2003. "Baculoviruses." In: *Biological Control: A Guide to Natural Enemies in North America.* Weeden, C.R., Shelton, A.M., Li, Y. and Hoffman, M.P. (eds). Cornell University, Ithaca, NY.

Danford, G. 1997. *Benecol: A revolutionary invention.* Retrieved May 1997 (Online Version: http://myy/helia.fi/~dange/raisio.html).

Dansons Group Inc. 2003. *Wood pellets.* Retrieved May 2004 from: http://www. dansons.com/wood-pellets/.

David Suzuki Foundation 2004. *Climate change: Impacts and solutions.* Retrieved Feb. 2004 from: http://www.davidsuzuki.org/Climate_Change/Solutions/Landfills.asp.

Davidson-Hunt, I., Duchesne, L.D. and Zasada, J.C. 2001. *Non-timber forest products: Local livelihoods and integrated forest management.* In: Forest Communities in the Third Millennium: Linking research, business and policy toward a sustainable non-timber forest product sector. (Online Version: http://www.ncrs.fs.fed.us/pubs/gtr/other/gtr-nc217/).

de Geus, P.M.J. 1995. *Botanical forest products in British Columbia: An overview.* B.C. Forest Service. Integrate Resources Policy Branch, SIL 457, 62 pp.

de Graaf, J., de Sauvage Nolting, P.R.W., van Dam, M., Belsey, E.M., Kastelein, J.J.P., Pritchard, P.H. and Stalenhoef, A.F.H. 2002. *Consumption of tall oil-derived phytosterols in a chocolate matrix significantly decreases plasma total and low-density lipoprotein-cholesterol levels.* British Journal of Nutrition (2002), 88:479-488.

Deluga, G.A., Salge, J.R., Schmidt, L.D. and Verykios, X.E. 2004. *Renewable hydrogen from ethanol by autothermal reforming.* Science 303:993-997.

Dey, D. 2001. *Commercial medicinal herb enterprise.* Agriculture Business Profiles, Jan 2001. Alberta Agriculture, Food and Rural Development. (Online Version: http://www.agric.gov.ab.ca/agdex/200/263_830-2.pdf).

Dominy, S.W.J. 2004. *Forest 2020 plantation demonstration and assessment initiative: An overview for Ontario.* Unpublished File Document. Natural Resources Canada, Canadian Forest Service, Great Lakes Forest Centre.

Doty, F.D. 2004. *A realistic look at hydrogen price projections.* Retrieved May 2004 from: http://www.dotynmr.com/PDF/Doty_H2Price.pdf.

Duchesne, L.C. and Weber, M.G. 1993. *High incidence of the edible morel morchella conica in a Jack pine, Pinus banksiana, forest following prescribed burning.* The Canadian Field-Naturalist 107(1):114-116.

Duchesne, L.C. and Davidson-Hunt, I. 1998. *Non-timber forest product exploitation in Canada.* In: Proceedings North American Forestry Association Annual Meeting.

Duchesne, L.C., Zasada, J.C. and Davidson-Hunt, I. 2000. *Nontimber forest product industry in Canada: Scope and research needs.* For. Chron. 76(5):743-746.

Duchesne, L.C. Zasada, J.C. and Davidson-Hunt, I. 2001. *Ecological and biological considerations for sustainable management of non-timber forest products in northern forests.* In: Forest Communities in the Third Millennium: Linking research, business and policy toward a sustainable non-timber forest product sector. (Online Version: http://www.ncrs.fs.fed.us/pubs/gtr/other/gtr-nc217/).

Duchesne, L.D., Davidson-Hunt, I. and Zasada, J.C. 2001a. *The future of the non-timber forest product industry. In:* Forest Communities in the Third Millennium: Linking research, business and policy toward a sustainable non-timber forest product sector. (Online Version: http://www.ncrs.fs.fed.us/pubs/gtr/other/gtr-nc217/).

DuPont 2003. *DuPont receives US EPA's presidential green chemistry award for new innovation clothing from cornfields: Bio-based process uses renewable resources instead of petrochemicals.* Retrieved June 2005 from: http://ca.dupont.com/NASApp/dupontglobal/corp/index.jsp?page=/content/US/en_US/news/releases/2003/nr06_24_03a.html.

DynaMotive 2005. *Corporate profile.* Retrieved May 2005 from: http://www.dynamotive.com/profile/.

Economist Intelligence Unit 2004. *World food and drinks: chew on this.* Retrieved July 2005 from: http://www.eiu.com/.

Eklund, R. and Pettersson, P. 2000. *Dilute-acid hydrolysis of softwood forest residue.* In: International Symposium on Alcohol Fuels XIII. Stockholm, Sweden.

Emery, M.R. 2001. *Non-timber forest products and livelihoods in Michigan's upper peninsula.* In: Forest Communities in the Third Millennium: Linking research, business and policy toward a sustainable non-timber forest product sector. (Online Version: http://www.ncrs.fs.fed.us/pubs/gtr/other/gtr-nc217/).

Emery, R.W. 2003. *Recycling the final residues of the paper industry.* Emery International Developments Ltd. 2003. Retrieved Dec. 2003 from: http://www.emeryinternational. com/newsviews/recycle.html.

Energex 2004. *Products: Premium grade residential pellet.* Energex Pellet Fuel Inc. (US). Granules Combustables Energex, Inc. (Canada). Retrieved Apr. 2004 from: http://www. energex.com/energex_contact.htm.

Energy Information Administration 1998. *Impacts of the Kyoto protocol on US energy markets and economic activity.* EIA Reports, SR/OIAF/98-03, Distribution Category UC-950. US Department of Energy. Washington, DC. (Online Version: http://www. eia.doe.gov/oiaf/kyoto/pdf/sroiaf9803.pdf).

Ensyn Group Inc. 2000. *The Ensyn technology – biomass applications.* Ensyn News and Information, Oct. 23, 2000. Retrieved May 2005 from: http://www.ensyn.com/info/ 23102000.htm.

Ensyn Group Inc. 2001. *The conversion of wood and other biomass to bio-oil.* ENSYN Group Inc., Boston MA. June, 2001. Retrieved May 2004 from: http://www.ensyn.com/info/ 01062001.pdf.

Environment Canada 1999. *The importance of nature to Canadians: Survey highlights.* Federal-Provincial-Territorial Task Force on the Importance of Nature to Canadians. 1999. Cat En 47-311/1999E. 50 pp.

Environment Canada 2000. *The importance of nature to Canadians: The economic significance of nature-related activities.* Federal-Provincial-Territorial Task Force on the Importance of Nature to Canadians. Cat En 47-312/2000E. 49 pp.

Environment Canada 2003. *State of the environment infobase, The state of Canada's environment - 1996.* National Indicators and Reporting Office, Hull, Quebec. (Online Version: http://www.ec.gc.ca/soer-ree/English/SOER/1996report/Doc/1-7-5-5-5-2-1.cfm).

Environment Canada 2005. *Government of Canada moves to create a market for emission reductions in all sectors of the economy.* News Release, Ottawa, August 11, 2005. (Online Version: www.ec.gc.ca/press/2005/050811_n_e.htm).

European Commission 1997. *Energy for the future: Renewable sources of energy, white paper for a community strategy and action plan.* Communication from the Commission, Nov. 26, 1997. (Online Version: http://europa.eu.int/comm/energy/library/599fi_en.pdf).

Expert Panel on the Commercialization of University Research 1999. *Public investments in university research: Reaping the benefits.* (Advisory Council on Science and Technology). Government of Canada, Ottawa, Canada.

Fahy, C. 1996. *BioMetics fuels growth with trash to gas plant construction.* Mass High Tech, Sept. 30 - Oct. 6, 1996. Volume 14; Issue 33. (Online Version: http://www.biometicsma. com/news.htm).

Falls Brook Centre 2003. *Balsam fir session.* In: Economic Opportunities in Non-timber forest Products. Atlantic region Non-timber Forest Products Workshop. Maritime Forest ranger School, Fredericton. Retrieved Dec. 2004 from: http://www.fallsbrookcentre. ca/webmain/programs/Forest/NTFP%20Web/econ_opp.htm.

Farnsworth, N.J. and Soejarto, D.D. 1985. *Potential consequence of plant extinction in the United States on the current and future availability of prescription drugs.* Economic Botany 39(3):231-240.

Fédération des producteurs acéricoles du Québec 2004. *Maple syrup: World production.* Retrieved Mar. 2004 from: http://www.maple-erable.qc.ca/f_sirop.html.

Feeny, P. 2004. *Seeking best practices in ecotourism, an industry in transition.* Nature Canada, Spring 2004. Retrieved Feb. 2005 from: http://www.cnf.ca/magazine/ spring04/ecologic.htm.

Finlayson, D. 2004. *Cure for common cold may be soon on the way.* Sault Star, Oct. 6, 2004.

Fitzpatrick, K. 1999 (ed.) *NutraNews.* Volume 3, June 1999. Saskatchewan Nutraceutical Network. Retrieved Sept. 2002 from: http://www.nutranet.org/subpages/newsletter/ nutranews3.html.

Fitzpatrick, K.C. 2002. *Canadian industry, research and regulatory trends in nutraceuticals, natural health products and functional foods.* Canadian Institute of Food Science and Technology. (Online Version: http://www.nffa.ca/NFFCanada.cfm).

Food and Agriculture Organization of the United Nations 2002. *Non-wood forest products from temperate broad-leaved trees.* ISBN 92-5-1048555-x. (Online Version: http:// www.fao.org/DOCREP/005/Y4351E/y4351e00.htm).

Food and Agriculture Organization of the United Nations 2005. *Agroforestry systems.* Retrieved June 2005 from: http://www.fao.org/forestry/foris/webview/forestry2/index.jsp? siteId=3162&sitetreeId=9470&langId=1&geoId=0.

Foragen 2005. *A vision for Canada making Canadians the healthiest people in the world.* Foragen Visions, Volume 3. Number 1 (May 2005), published for Foragen Management Technologies Inc. by The Signature Group, Saskatchewan. Retrieved July 2005 from: http://www.foragen.com/html/newsletters/visions3_1.pdf.

Forbes Medi-Tech Inc. 2002. *Construction of manufacturing plant completed.* Press release. Jan. 15, 2002. Retrieved Jan. 2003 from: http://www.siliconinvestor.com/stocktalk/msg. gsp?msgid=16911507.

Forbes Medi-Tech. Inc. 2003. *Pharmaceuticals.* Retrieved Jan. 2003 from: http://www. forbesmedi.com/s/Home.asp.

Forbes Medi-Tech Inc. 2004. *Nutraceuticals - Over-the-counter dietary supplements.* Retrieved July 2005 from: http://www.forbesmedi.com/s/DietarySupplements.asp.

Forbes Medi-Tech Inc. 2004a. *Forbes Medi-Tech announces financial results for the second quarter ended June 30, 2004.* News release. Retrieved Nov. 2004 from: http://www. forbesmedi.com/s/News-2004.

Forest Products Association of Canada 2005. *The Canadian forest. Forest management: forest renewal.* Retrieved Sept. 2005 from: http://www.fpac.ca/en/sustainability/index.php

Forest Products Association of Canada 2005a. *Current issues. Embracing bioenergy.* Retrieved Sept. 2005 from: http://www.fpac.ca/en/sustainability/current_issues.php.

Foster, S. 1995. *Forest pharmacy, medicinal plants in American forests.* Forest History Society, Durham, North Carolina, 57 pp.

Freel, B., Graham, R. 2000. *Commercial bio-oil production via rapid thermal processing.* Ensyn Group Inc., Boston MA. Retrieved May 2004 from: http://www.ensyn.com/ info/11122000.htm.

Freeman, S. 1997. *An estimate of pine mushroom productivity in the Nahatlatch watershed.* Forest Renewal BC Report HQ96174-HE.

Friends of Clayoquot Sound 2004. *Clayoquot green economic opportunities project: ecotourism sector analysis.* Retrieved Feb. 2005 from: http://www.focs.ca/reports/ cgeo2_4.html.

Fur Council of Canada 2004. *Canadian fur trade at a glance.* Retrieved Feb. 2005 from: www.furcouncil.com.

Fytokem 2002. *Fytokem announces positive test results-Canadian Willowherb^{TM}. Proven effective in killing acne bacterium.* Press Release, Saskatoon, Saskatchewan, June 18, 2002. Retrieved Nov. 2005 from: http://www.fytokem.com/news/PR-2002-06-18.pdf.

Fytokem 2004. *Canadian Willowherb^{TM}.* Retrieved Apr. 2004 from: http://www. fytokem.com/products/canwillowherb.php.

Gagnon, D. 1999. *A review of the ecology and population biology of Goldenseal, and protocols for monitoring its populations.* Medicinal Plant Working Group. Green

Medicine. Retrieved Apr. 2005 from: http://www.nps.gov/plants/medicinal/pubs/goldenseal.htm.

Garry, V.I., Tarone, R.E., Kirsch, I.R., Abdallah, J.M., Lombardi, D.P., Long, L.K., Burroughs, B.L., Barr, D.B. and Kesner, J.S. 2001. *Biomarker correlations of urinary 2,4-D levels in foresters: Genomic instability and endocrine disruption.* Env. Health Perspectives 109:495-500.

Genome Canada 2005. *What's new at genome Canada?* Retrieved Aug. 2005 from: http://www.genomecanada.ca/home.asp?l=e.

Ghorpade, V.M. and Hanna, M.A. 1997. "Industrial applications for levulinic acid". In: *Cereals - Novel Uses and Processes.* Chapter 7, pp. 49-55. Campbell, G.M., Webb, C. and McKee, S.L. (eds). Plenum Press, New York, NY.

Globerman, S. and Vertinsky, I.B., 1995. *Forest biotechnology in Canada: Analysis of intellectual property rights and protection of higher lifeforms.* Intellectual Property Policy Directorate, Industry Canada. 126 pp. Retrieved Feb. 2003 from: http://strategis.ic.gc.ca/pics/ip/glberef.pdf.

Gordon, A.M. and Newman, S.M. (Eds.) 1997. *Temperate agroforestry systems.* CAB Intenational, UK. ISBN 0 85199 147 5.

Government of Canada 1999. *National climate change process. Sinks table options paper: land use, land use change and forestry in Canada and the Kyoto protocol.* Canadian Forest Service, Natural Resources Canada, and Environment Canada. 165 p + appendices. (Online Version: http://www.ec.gc.ca/pdb/ghg/documents/Optionspaper.pdf).

Government of Canada. 2000. *Government of Canada action plan 2000 on climate change.* (Online Version: http://climatechange.gc.ca/english/publications/ap2000/ Action_Plan_2000_en.pdf).

Government of Canada 2002. *The functional food and nutraceutical industry: innovation profile.* July 2002. Retrieved Apr. 2004 from: http://www.innovation.gc.ca/gol/innovation/interface.nsf/vSSGBasic/in02585e.htm.

Government of Canada 2002a. *The Canadian pharmaceutical industry. Canada's innovation strategy.* Sector Profiles April 2002, Innovation Profile. (Online Version: http://www.innovationstrategy.gc.ca/cmb/innovation.nsf/SectoralE/Pharmaceuticals).

Government of Canada 2002b. *Canada's innovation strategy.* Retrieved Nov. 2005 from: http://innovation.gc.ca/gol/innovation/site.nsf/en/in04113.html.

Government of Canada 2005. *Project green. Moving forward on climate change: a plan for honouring our Kyoto commitment.* (Online Version: http://www.climatechange.gc.ca/english/newsroom/2005/plan05.asp).

Granit 2005. *Renewable materials, Lignin polymer.* Retrieved June 2005 from: http://www.granit.net/index.htm.

Greene, N. 2004. *Growing energy: How biofuels can help end America's oil dependence.* The Natural Resources Defense Council. Retrieved Nov. 2005 from: http://www.bioproducts-bioenergy.gov/pdfs/NRDC-Growing-Energy-Final.3.pdf.

Gregg, D.J. and Saddler, J.N. 1996. *A techno-economic assessment of the pretreatment and fractionation steps of a biomass-to-ethanol process.* Appl. Biochem. Biotechnol. 57/58, 711-727.

Gregg, D.J., Boussaida, A. and Saddler, J.N. 1998. *Techno-economic evaluations of a generic wood-to-ethanol process: Eeffect of increased cellulose yields and enzyme recycle bioresource technology.* Bioresource Technol. 63:7-12.

Griss, P. 2002. *Forest carbon management in Canada.* Final report of the Pollution Probe Forest Carbon Management Series. (Online Version: www.pollutionprobe.org/whatwedo/FCM/finalreport.pdf).

Griss, P. 2004. Forest carbon management - overview. In: *Pollution Probe Forest Carbon Management (FCM) Pilots Series, FCM Protocol Template and Guidance for Project Development July 14, 2004.*

Hakkila, P. 2000. *Forest chips in Finland - use, experiences and prices.* OPET Finland Energy Jyvaskiyla . Jyvaskiyla, Finland, pp. 39-55.

Hall, J.P. 2001. *Bioenergy in Canada 2001.* Bioenergy Task 31 News. Task 31: Conventional Forestry Systems for Sustainable Production of Bioenergy. Vaxjo Sweden.

Hanam Canada Corporation 2004. *Canadian market for wood residue.* Retrieved Mar. 2005 from: http://www.islandnet.com/~hburke/sawmill.htm.

Hansen, E. 1992. *Mid-rotation yields of biomass plantations in the north central United States.* United States Department of Agriculture, Forest Service, North Central Forest Experiment Station, Research Paper NC-309. (Online Version: http://www.ncrs.fs.fed.us/pubs/rp/rp_nc309.pdf).

Hardin, B. 2000. *Making xylitol sweetener from corn.* Agriculture Research magazine, July 2000. (Online Version: http://www.ars.usda.gov/is/AR/archive/jul00/xylit0700.htm).

He, S-A. and Sheng, N. 1995. *"Utilization and conservation of medicinal plants in China, with special reference to Atractyoides lancea."* In: Non-Wood Forest Products 11, Medicinal Plants for Conservation and Health Care. Food and Agriculture Organization of the United Nations, 8 pp. (Online Version: http://www.fao.org/docrep/W7261e/W7261e13.HTM).

Health Canada 1998. *Nutraceuticals/functional foods and health claims on foods.* Policy paper. Therapeutic Products Programme and the Food Directorate from the Health Protection Branch, Nov. 2, 1998. (Online Version: http://www.hc-sc.gc.ca/food-aliment/ns-sc/ne-en/health_claims-allegations_sante/pdf/e_nutra-funct_foods.pdf).

Health Canada 2003. *Natural health products regulations.* 2003-06-18 Canada Gazette Part II, Vol. 137, No. 13. (Online Version: http://www.hc-sc.gc.ca/hpfb-dgpsa/nhpd-dpsn/regs_cg2.pdf).

Hester, G. 2004. *The honeybee. Teachers' kit.* The Ontario Beekeepers' Association. Retreived Apr. 2005 from: http://www.ontariobee.com/4_resources/teachers.htm.

Hobbs, J.E. 2001. *Developing supply chains for nutraceuticals and functional foods: opportunities and challenges.* Institute of Nutraceutical and Functional Foods, Centre for Research in the Economics of Agrifood. Research Series SR.01.05 (Part 4), 18 pp.

Houghton, J.T., Ding, Y., Griggs, D.J., Noguer, M. van der Linden, P.J., Dai, X., Maskell, K., Johnson, C.A. (eds) 2001. *Climate change 2001: the scientific basis.* Cambridge University Press, Cambridge, U.K.

Howatt, K. 1990. *Azadirachta indica: One tree's arsenal against pests.* Colorado State University. Retrieved Aug. 2005 from: http://www.colostate.edu/Depts/Entomology/courses/en570/papers_1994/sclar.html.

Howe Sound Pulp and Paper Ltd. 2000. *1999 Emissions of greenhouse gases: A progress report.* Participation in the Canadian Voluntary Challenge Registry. July 14, 2000 (Online Version: http://www.vcr-mvr.ca/registry/out/C3031-HOWE-PDF.PDF).

Huang, Y.G., Branka, B. and Ivanochko, G. 1999. *Selected non-timber forest products with medicinal applications from Jilin province in China.* In: Forest Communities in the Third Millenium: Linking research, business, and policy toward a sustainable non-timber forest product sector. (Online Version: http://www.ncrs.fs.fed.us/pubs/gtr/other/gtr-nc217/).

Human Resources Development Canada 1998. *The changing skill structure of employment in Canada.* R-99-7E. Applied Research Branch. Strategic Policy. 1998. ISBN: 0-662-27913-1. Retrieved Sept. 2005 from: http://www11.sdc.gc.ca/en/cs/sp/arb/publications/research/1998-001335/1998-001335.pdf.

Hutton, A. 2002. *Officials admit mistake in fire at dump.* Kansas State Firefighters Association. Olathe Daily News, July 4, 2002. Retrieved May 2003 from: http://www.ksffa.com/.

Hyvonen, R., Olsson, B.A., Lundkvist, H. and Staaf, H. 2000. *Decomposition and nutrient release from Picea abies* (L.) *Karst. and Pinus sylvestris L. logging residues.* Forest Ecology and Management 126:97-112.

Ihalainen, M., Alho, J., Kolehmainen, O. and Pukkala, T. 2002. *Expert models for bilberry and cowberry yields in Finnish forests.* Forest Ecology and Management 157:15-22.

IHB Timber Exchange 2004. *Press releases.* Retrieved Mar. 2004 from: http://www.timber-exchange.com/.

Immunoceuticals Inc. 2004. *Mushroom mycelium products.* Retrieved Oct. 2004 from: http://www.immunoceuticals.nwbotanicals.org/.

IMS Health 2003. *2002 World pharma sales growth: slower, but still healthy.* IMS Health Online Store. Retrieved Mar. 2003 from: http://open.imshealth.com/webshop2/IMSinclude/i_article_20030228.asp.

Industry Canada 2002. *Innovation in Canada. The Canadian forest industry.* Retrieved Sept. 2005 from: http://strategis.ic.gc.ca/epic/internet/infi-if.nsf/en/h_fb01336e

Industry Canada 2004. *Innovation roadmap on bio-based feedstocks, fuels and industrial products: capturing Canada's natural advantage.* Industry Canada. Manufacturing Industries Branch. Cat. No. Iu44-11/2004?E_HTML. ISBN 0-662-36411-2. (Online Version: http://www.bio-productscanada.org/pdf/en_roadmap_book.pdf).

Industry Canada 2004a. *Trade data online.* Retrieved Dec. 2004 from: http://strategis.ic.gc.ca/sc_mrkti/tdst/tdo/tdo.php.

Industry Canada 2005. *Trade data online. Canadian and US trade by product (HS codes).* Data from Statistics Canada and the US Census Bureau. Retrieved Apr. 2005 from: http://strategis.ic.gc.ca/sc_mrkti/tdst/engdoc/tr_homep.html.

Industry Canada 2005a. *Biostrategy.* Retrieved Aug. 2005 from: http://strategis.ic.gc.ca/utils/htmlMessages/redirectCSB-SCBe.html#3b.

Institute On Governance 2001. *A primer for scientists: Ethical issues of environmental biotechnology research.* Ottawa, Canada. Retrieved Sept. 2005 from: http://www.iog.ca.

International Diabetes Federation 2003 *Did you know?* Retrieved Feb. 2003 from: http://www.idf.org/home.

International Emissions Trading Association 2002. *GHG trading market.* Retrieved Oct. 2005 from: www.ieta.org.

International Energy Association 2005. *The Charlottetown district energy system.* Task 29: IEA bioenergy network on Socio-economics. Retrieved May 2005 from: http://www.aboutbioenergy.info/Charlottetown.html.

International Fair Trade Association 2004. *What is fair trade?* Retrieved June 2005 from: http://www.ifat.org/whatisft.html.

Iogen 2004. *Cellulose ethanol is ready to go.* News release April 21, 2004. Retrieved Apr. 2004 from: www.iogen.ca.

Japan External Trade Organization 2004. *Japan xxternal trade organization.* Retrieved Apr. 2005 from: http://www.jetro.go/.

Jasinski, S.M. 2000. *Peat.* US Geological Survey, Mineral Commodity Summaries, Feb. 2000. Retrieved Feb. 2004 from: http://minerals.usgs.gov/minerals/pubs/commodity/peat/510300.pdf.

Jones, P.J.H., Fady, Y.N., Mahmouh, R.S. and Vanstone, C.A. 1999. *Cholesterol-lowering efficacy of a sitostanol-containing phytosterols mixture with a prudent diet in hyperlipidemic men.* Am. J. Clin. Nutr. 69:1144-1150.

JUCA Super Fireplace 2003. *The amount of energy in wood fuel.* Technical series, JUCA Super Fireplace Products. Retrieved Feb. 2003 from: http://mb-soft.com/juca/print/311.html.

Kaegi, E. 1998. *Unconventional therapies for cancer: 1. Essiac.* Canadian Medial Ass. J. 158(7):897-902.

Kalt, W., McDonald, J.E., Ricker, R.D. and Lu, X. 1999. *Anthocyanin content and profile within and among blueberry species.* Can. J. Plant Sci. 79:617-623.

Kaufman, P.B., Cseke, L.J., Warber, S., Duke, J.A. and Brielmann, H.L. 1998. *Natural products from plants.* CRC Press, New York, NY. 343 pp.

Kerns, B.K., Liegel, L., Pilz, D. and Alexander, S.J. 2002. "Biological Inventory and Monitoring." In: *Nontimber Forest Products in the United States.* Jones, E.T., McLain, R.J. and Weigland, J. (eds.). University Press of Kansas, Lawrence, K.S. pp. 115-150.

Kettle, R. 2001. *Promising future for bioenergy in the United Kingdom.* IEA Bioenergy Library. Bioenergy News Vol 13 #1 June 2001. Retrieved May 2005 from: http://www.ieabioenergy.com/library/77_v13no1.pdf.

Krantz, J. 2001. *The Minnesota approach to non-timber forest product marketing: the balsam bough industry and other examples.* In: Forest Communities in the Third Millennium: Linking research, business and policy toward a sustainable non-timber forest product sector. (Online Version: http://www.ncrs.fs.fed.us/pubs/gtr/other/gtr-nc217/).

Krigmont, H.V. 1999. *Integrated biomass gasification combined cycle (IBGCC) power generation concept: The gateway to a cleaner future.* A white paper. Allied Environmental Technologies, Inc. Retrieved Apr. 2004 from: http://www.alentecinc.com/papers/IGCC/ADVGASIFICATION.pdf.

Kuipers, S.E. 1995. *Trade in medicinal plants.* Food and Agriculture Organization of the United Nations, 17, pp. (Online Version: http://www.fao.org/docrep/W7261e/W7261e08.HTM).

Langlais, G. 2003. (draft). *Aménagement d'érablières en vue de favoriser la production de plantes à valeur ajoutée (Rapport préliminaire-Le rapport final sera disponible dès février 2005) (Practical current and future prospects of agroforestry in maple stands.)* Groupe de recherche appliquée en agroforesterie (GRAAF). (Online Version: http://www.graaf.ca/cultur.html).

Larex Inc. 2003. *Natural, bioactive compounds from the chemistry of trees.* Retrieved Apr. 2004 from: http://www.larex.com/htm/company.html.

Latta, S.A. 1999. *Herbs: feverfew.* Kansas State University Agricultural Experiment Station and Cooperative Extension Service. Kansas State University. Retrieved Apr. 2004 from: http://www.oznet.ksu.edu/library/fntr2/mf2379.pdf.

Lazar, A. 2005. *A proud record of leadership in addressing and adapting to climate change.* For Chron 81(5):631-633.

Lehtikangas, P. 2001. *Quality properties of pelletised sawdust, logging residues and bark.* Biomass and Bioenergy 20:351-360.

Lenox Polymers Ltd. 2000. *Lenox polymers Ltd.* Retrieved August 2004 from: http://www.lenoxpoly.com/.

Letchworth, B. 2001. *The industry of wildcrafting, gathering, and harvesting of NTFPs: An insider's perspective.* In: Forest Communities in the Third Millennium: Linking research, business and policy toward a sustainable non-timber forest product sector. (Online Version: http://www.ncrs.fs.fed.us/pubs/gtr/other/gtr-nc217/).

Leuty, T. 1999. *Walnuts for food versus walnuts for wood.* Ontario Ministry of Agriculture and Food. Retrieved Jan. 2003 from: http://www.gov.on.ca/OMAFRA/english/crops/facts/info_walnuts_food.htm.

Lignin Institute 1992. *Lignins–products with many uses.* Retrieved June 2005 from: http://www.lignin.info/.

Lipinsky, E.S. 1981. *Chemicals from biomass: petrochemical substitution options.* Science 212:1465-1471.

Lizotte, P.A. and Knapp, G.E. 2003. *Cancer cure or conservation: A question of health for humans and the ecosystem.* National Center for Case Study Teaching in Science, University at Buffalo, State University of New York. Retrieved June 2005 from: http://www.sciencecases.org/taxol/taxol.pdf.

Lowe, J.J., Power, K. and Gray, S.L. 1996. *Canada's forest inventory 1991: the 1994 version.* An addendum to Canada's forest inventory 1991. Natural Resources Canada, Canadian Forest Service, Pacific Forestry Centre, Victoria, BC. Information Report BC-X-362E.

Lulsdorf, M.M., Tautorus, T.E., Kikcio, S.I., Bethune, T.D. and Dunstan, D.I. 1993. *Germination of encapsulated embryos of interior spruce (Picea glauca engelmannii complex) and black spruce (Picea mariana Mill.).* Plant Cell Rep. 385-389.

Manitoba Agriculture, Food and Rural Initiatives 2001. *Herb & spice industry overview - nutraceuticals - functional foods / opportunities for Manitoba producers in the herb and spice industry.* Retrieved Apr. 2004 from: http://www.gov.mb.ca/agriculture/financial/agribus/ccg02s05.html.

Manitoba Agriculture, Food and Rural Initiatives 2001a. *Evening primrose.* Retrieved Mar. 2004 from: http://www.gov.mb.ca/agriculture/crops/medicinal/bkq00s04.html.

Marles, R.J. 1996. *Perspectives on medicinal uses of native plants.* Prairie Medicinal and Aromatic Plants Conference, Olds, AB. March 3-5, 1996. (Online Version: http://www.agric.gov.ab.ca/crops/special/medconf/marles.html).

Marles, R.J. 2001. *Non-timber forest products and Aboriginal traditional knowledge.* In: Forest Communities in the Third Millennium: Linking research, business and policy toward a sustainable non-timber forest product sector. (Online Version: http://www.ncrs.fs.fed.us/pubs/gtr/other/gtr-nc217/).

Masse, S. 2005. *Developing short-rotation willow plantation/agroforestry systems for bioenergy generation in Canada.* Approved project for Canadian Biomass Innovation Network. (Online Version: http://www.cbin.gc.ca/Abstracts2004-e.html).

Mauser, G. 2005. *Hunters are the mainstay of provincial wildlife management programs.* Retrieved Feb. 2005 from: http://www.sfu.ca/~mauser/papers/hunters/hunt-licence-rept.pdf.

Maynard, A.A. and Hill, D.E. 1992. *Vegetative stabilization of logging roads and skid trails.* North. J. Appl. For. 9(4):153-157.

McCallum, B. 1999. *Woodchip supply system options for remote communities.* Natural Resources Canada, Canadian Forest Service, Great Lakes Forestry Centre. Infor. Rep. GLC-X-3. 19 p.

McCormick, B. 2003. *Renewable diesel fuels: Status of technology and R&D needs.* In: 8th Diesel Engine Emissions Reduction Conference. National Renewable Energy Laboratory.

McIssac, D.W. 1993. *Sugar bush establishment and management.* Nova Scotia Soils Institute. Proceedings of the Agroforestry Workshop March 29-30, 1993. Truro, NS. (Online Version: http://www.gov.ns.ca/nsaf/nssi/workshops/macissac.htm).

McKenney, D.W., Yemshanov, D., Fox, G. and Ramlal, E. 2004. *Cost estimates for carbon sequestration from fast growing poplar plantations in Canada.* Forest Policy and Economics 6 (3-4):345-358.

Mehaffey, J., Richardson, L., Batista, M. and Gueorguiv, S. 2000. *Self-heating and spontaneous ignition of fibreboard insulating panels.* Fire Technology 36:4:2000.

Mendelsohn, R. and Balick, M.J. 1995. *The value of undiscovered pharmaceuticals in tropical forests*. Econ. Bot. 49:223-228.

Miner, R. 2005. *Biomass carbon accounting from national and corporate perspectives: an overview of national inventory accounting*. Power Point Presentation to the Forest Products Association of Canada Climate Change Committee, Feb. 16, 2005.

Ministère des Ressources Naturelles 2002. Rapport du comité interministériel sur la contribution des terres du domaine de l'état, au développement de l'industrie du bleuet. (Online Version: http://www.mrnfp.gouv.qc.ca/publications/forets/rapport-bleuets.pdf).

Mitchell, M. and Associates 1997. *The harvest, market and availability of special forest products in the Manitoba model forest*. The Manitoba Model Forest, Pine Falls Manitoba. Project: 95-4-09.

Mitchell, M., Tyshenko, M. and Leiss, W. 1998. *Environmental release of transgenic trees in Canada – potential risks and benefits*. A Canadian Forest Service discussion paper for Natural Resources Canada.

Miyamoto, K. (ed.) 1997. *Renewable biological systems for alternative sustainable energy production*. FAO Agricultural Services Bulletin 128. Retrieved Feb. 2003 from: http://www.fao.org/docrep/w7241e0j.htm.

Mohammed, G.H. 1999. *Non-timber forest products in Ontario: an overview*. Forest Research Information Paper No. 145, Ontario Forest Research Institute, Ontario Ministry of Natural Resources. 71 pp. (Online Version: http://www.mnr.gov.on.ca/MNR/forests/t&t_research/publications/145.pdf).

Mohammed, G.H. 2001. *Recommendations of sustainable development of non-timber forest products*. In: Forest Communities in the Third Millennium: Linking research, business and policy toward a sustainable non-timber forest product sector. (Online Version: http://www.ncrs.fs.fed.us/pubs/gtr/other/gtr-nc217/).

Montgomery, T. 2000. *Native crops project for Hull's field. Using native plants to link local agriculture initiatives with environmental stewardship*. Urban Agriculture Notes, July 2000, Published by City Farmer, Canada's Office of Urban Agriculture. (Online Version: http://www.cityfarmer.org/nativecrops.html).

Moore, T. 1996. *Harvesting the benefits of biomass*. EPRI Journal 3:16-25.

Morris, D. 2003. *Reflections of a hydrogen economy*. The Carbohydrate Economy Newsletter, Vol. 5(1), Winter 2003. Retrieved Apr. 2004 from: http://www.carbohydrateeconomy.org/

Munson, A.D., Samson, J. and Piche, Y. 1995. "Management of root symbioses in revegetation." In: *Research in Canada. Recent Progress in Forest Biotechnology in Canada*. PI-X-120. 65-73.

Mycova[SM] 2005. *Helping the ecosystem through mushroom cultivation*. Fungi Perfecti LLC. Retrieved Aug. 2005 from: http://www.fungi.com/mycotech/mycova.html.

Nadeau, I., Olivier, A., Simard, R. R., Coulombe, J., Nadeau, I. and Yelle, S. 1999. *Growing American ginseng in maple forests as an alternative land-use system in Quebec, Canada*. Agroforestry Systems. 44 (2-3):345-353.

Nath, K. and Das, D. 2003. *Hydrogen from biomass*. Current Science 85(3):265-271.

National Academy of Engineering and Board on Energy and Environmental Systems 2004. *The hydrogen economy: Opportunities, costs, barriers, and R&D needs*. National Academy of Engineering (NAE), Board on Energy and Environmental Systems (BEES). Retrieved May 2004 from: http://books.nap.edu/books/0309091632/html/378.html.

National Renewable Energy Laboratory 1998. *Life cycle inventory of biodiesel and petroleum diesel for use in an urban bus*. A joint study sponsored by: US Department of Agriculture and US Department of Energy. NREL SR-580-24089 UC Category 1503. Retrieved

May 2005 from: http://www.biodiesel.org/resources/reportsdatabase/reports/gen/19980501_gen-339.pdf.

National Round Table on Environment and Economy 2002. *The ABC's of emissions trading: an awareness-raising initiative.* Final Summary Report, Nov. 2001-Mar. 2002.

Natural Resources Canada 1998. *Report to the minister of natural resources Canada.* National Advisory Board on Forest Research.

Natural Resources Canada 2000. *Canadian forest sector production - 2000.* Retrieved May 2005 from: http://www.pfc.cfs.nrcan.gc.ca/canforest/canf/sector4_e.html.

Natural Resources Canada 2000a. *Canada's energy markets: Sources, transformation and infrastructure.* Energy in Canada 2000. Retrieved May 2004 from http://www2.nrcan.gc.ca/es/ener2000/omline/html/chap3b_e.cfm.

Natural Resources Canada 2002. *Renewable energy in Canada, status report 2002.* Prepared for the Office of Energy Research and Development. Retrieved May 2005 from: http://www2.nrcan.gc.ca/es/oerd/english/view.asp?x=700.

Natural Resources Canada 2003. *The state of Canada's forests 2002-2003.* (Online Version: http://www.nrcan.gc.ca/cfs-scf/sof/).

Natural Resources Canada 2003a. *Roundwood production by province, category and species group, 1940-2000.* Sources: Statistics Canada Cat. No. 25-201 and Compendium of Canadian Forestry Statistics 1998, National Forestry Database. Last revision: Oct. 2003. Retrieved Feb. 2004 from: http://mmsd1.mms.nrcan.gc.ca/forest/members/section1/I-2print.asp?lang=en&prv=20.

Natural Resources Canada 2003b. *Ethanol the road to a greener future.* Office of Energy Efficiency, Natural Resources Canada. Publication Number M92-257-2003. (Online Version: http://oee.nrcan.gc.ca/vehiclefuels).

Natural Resources Canada. 2003c. *Ground hemlock or eastern yew: A non-timber forest resource.* (Online Version. http://www.atl.cfs.nrcan.gc.ca/index-e/what-e/science-e/biotechnology-e/groundhemlock-e.html).

Natural Resources Canada 2004. *Energy use data handbook, 1990 and 1996 to 2002.* June 2004. Cat. No. M141-2/2002E. ISBN 0-662-36478-3. (Online Version: http://oee.nrcan.gc.ca/neud/dpa/data_e/Handbook04/Datahandbook2004.pdf).

Natural Resources Canada 2004a. *The state of Canada's forests 2003-2004.* Canadian Forest Service. Retrieved Nov. 2005 from: http://www.nrcan-rncan.gc.ca/cfs-scf/national/what-quoi/sof/sof04/brief_e.html.

Natural Resources Canada 2005. Forest protection using biological pest control methods. Canadian Forest Service. Retrieved Aug. 2005 from: http://www.nrcan-rncan.gc.ca/cfs-scf/science/biotechnology/forest_e.html.

Natural Resources Canada 2005a. Biotechnology at the Canadian forest service. Canadian Forest Service. Retrieved Aug. 2005 from: http://www.nrcan-rncan.gc.ca/cfs-scf/science/biotechnology/index_e.html.

Natural Resources Canada 2005b. *Statistics on natural resources. Facts on forestry.* Retrieved Sept. 2005 from: http://www.nrcan.gc.ca/statistics/forestry/default.html.

Network for Alternative Technology and Technology Assessment 2000. *Energy: a beginners guide.* Renew On Line (UK). Energy and Environment Research Unit, The open university, Milton Keynes. (Online Version: http://www-tec.open.ac.uk/eeru/natta/energy.html - 5).

New Internationalist 2000. *The facts on pesticides.* New Internationalist 323. May 2000. Retrieved Aug. 2005 from: http://www.newint.org/.

Newmaster, S.G., Lehala, A., Uhlig, P.W.C., McMurray, S. and Oldham, M.J. 1998. *Ontario plant list.* Ont. Min. Nat. Resour., Ont. For. Res. Inst., Sault Ste. Marie, On. For. Res. Info. Pap. No. 123.

Norbeck, J. and Johnson, K. 2000. *Evaluation of a process to convert biomass to methanol fuel.* US Environmental Protection Agency. EPA/600/SR-00/092.

Northwest Botanicals 2004. *Northwest Botanicals Inc.* Retrieved Apr. 2005 from: http://www.nwbotanicals.org/nwb/mushrooms.htm.

Odwalla 2001. *Odwalla introduces glorious morning: a glorious orange cranberry juice blend featuring larch arabinogalactan and calcium.* Press release. Retrieved Oct. 2004 from: http://www.odwalla.com/enwfiles/relase60.html.

Olivier, A. 2001. La protection et la mise en valeur des resources du milieu rural par l'agroforesterie. Colloque sur l'agroforesterie au Quebec. (Online Version: http://www.sbf.ulaval.ca/colloque-agf-2001/resum.html).

Ontario Ministry of Agriculture and Food 2002. *State of the industry, 2002 agroforestry in Canada.* Retrieved Apr. 2005 from: http://www.gov.on.ca/OMAFRA/english/crops/facts/info_state2001.htm.

Organic Crop Improvement Association International 2005. *Organic certification.* Retrieved Nov. 2005 from: http://www.ocia.org/.

Organisation for Economic Co-operation and Development 2003. *Tax incentives for research and development: trends and issues.* (Online Version: http://www.oecd.org/dataoecd/12/27/2498389.pdf).

Parsons, T.J., Sinkar, V.D., Stettler, R.F., Nester, E.W. and Gordon, M.P. 1986. *Transformation of poplar by Agrobacterium tumefaciens.* Bio/technol 4:533-536.

Pedlar, J.H., Pearce, J.L., Venier, L.A. and McKenney, D.W 2002. *Coarse woody debris in relation to disturbance and forest type in boreal Canada.* Forest Ecology and Management 158:189-194.

Penner, M., Power, K., Muhairwe, C., Tellier, R. and Wang, Y 1997. *Canada's forest biomass resources: Deriving estimates from Canada's forest inventory.* Pacific Forestry Centre, Canadian Forest Service, Victoria, BC, Information Report BC-X-370. (Online Version: http://warehouse.pfc.forestry.ca/pfc/4775.pdf).

Pest Management Regulatory Agency 2005. *Pest management regulatory agency.* Health Canada. Retrieved Aug. 2005 from: http://www.pmra-arla.gc.ca/english/index-e.html.

Peterson, C.L., Thompson, J.C. and Taberski, J.S. 1999. *One-thousand-hour engine durability test with HySEE and using a 5X-EMA test cycle.* Am. Soc. Agric. Eng. 42:23-30.

Phytogen Life Sciences Inc. 2002. *Corporate highlights.* Retrieved Mar. 2004 from: http://www.phytogen.com/strategy.htm.

Pitcher, K. 2000. *Turning willow into megawatts.* Renewable Energy World, Nov.-Dec. 2000, Volume 3 Number 6. (Online Version: http://www.jxj.com/magsandj/rew/2000_06/turninig_willow.html).

Pollution Probe 2003. *Emissions trading primer.* (Online Version: www.pollutionprobe.org/Publications/emissionstradingprimer.pdf).

Pollution Probe and BIOCAP Canada 2004. *Primer on bioproducts.* ISBN 0-919764-57-6. Retrieved Aug. 2005 from: http://www.biocap.ca/images/pdfs/BioproductsPrimerE.pdf.

Porter, B. and Barl, B. 2000. *Astragalus in Saskatchewan.* Saskatchewan Agriculture, Food and Rural Revitalization (Online Version: http://www.agr.gov.sk.ca/DOCS/crops/special_crops/astragalus0007.asp).

Porter, B. and Brenzil, C. 2003. *Farm facts: dandelion production.* Saskatchewan Agriculture, Food and Rural Revitalization. Retrieved Mar. 2004 from: http://www.agr.gov.sk.ca/docs/crops/special_crops/production_information/Dandelion.pdf.

Porter, B., Barl, B. and Kehler, C. 2000. Feverfew in Saskatchewan. Saskatchewan Agriculture, Food and Rural Revitalization (Online Version: http://www.agr.gov.sk.ca/crops/special_crops/feverfew.asp).

Powell, M. 2001. *Producing safe and quality products from the land.* Retrieved Dec. 2004 from: http://www.elements.nb.ca/theme/ethnobotany/wreaths/wreaths.htm.

Powers, R.F. 1999. *On the sustainable productivity of planted forests.* New Forests 17:263-306.

Presusser, S. 2004. *Berlin S&T report 37/2004: BtL fuels symposium.* Technology Officer, Canadian Embassy.

Priesnitz, W. 2000. *Government committee calls for end to cosmetic pesticides.* Natural Life Magazine: 74: 2000. Life Media.

Primary Power International 2005. *Creating energy today for tomorrow's needs.* Retrieved Feb. 2005 from: http://www.primarypower.com/index.htm.

Prime Minister's Office (Finland) 2000. *Environmental and energy taxation in Finland - Preparing for the Kyoto challenge.* Summary of the Working Group Report. Prime Minister's Office Publication Series 2000/4. Helsinki, Finland. (Online Version: http://www.vn.fi/vnk/english/publications/vnk20004e/publication20004.pdf).

Prince Edward Island Department of Environment, Energy and Forestry 2002. *Ground hemlock harvesting guidelines.* Developed by Natural Resources Canada, Canadian Forest Service, Fredericton, and the Prince Edward Island Department of Environment, Energy and Forestry. Retrieved Jan. 2005 from: http://www.gov.pe.ca/photos/original/af_hemlock0402.pdf.

Rafaschieri, A., Rapaccini, M. and Manfrida, G. 1999. *Life cycle assessment of electricity production from poplar energy crops compared with conventional fossil fuels.* Energy Cons. Manage. 40 (1999): 1477-1493. (Online Version: http://doi.eng.cmu.ac.th/Thailca/pdf/LCA_of_electricity_production.pdf).

Regan, S. and Rutledge, B. 1995. "Molecular research into conifer growth and development." In: *Research in Canada. Recent Progress in Forest Biotechnology in Canada.* PI-X-120. 42 51.

Regional Wood Energy Development Program 2002. *Biomass briquetting: technology and practices.* Retrieved Feb. 2005 from: http://www.rwedp.org/fdch1.html.

Reichenbach de Sousa, L.C. 2001. *Gasification of wood, urban wastewood (Altholz) and other wastes in a fluidised bed reactor.* A disseration submited to the Swiss Federal Institute of Technology Zürich, Switzerland. (Online Version: http://e-collection.ethbib.ethz.ch/ecol-pool/diss/abstracts/p14207.pdf).

Renewable Oil International 2005. *Biooil as a means of energy densification.* Retrieved June. 2005 from: http://www.renewableoil.com/.

Renzie, C. and Han H-S. 2002. *Operational analysis of partial cut and clearcut harvesting methods in north-central interior ICH mountain forests - part 2.* Final Report to the Robson Valley Enhanced Forest Management Pilot Project (EFMPP), Mar. 2002.

Richters Herbs 2002. *What happened to the herbal markets?* Miller R.A. (ed). Retrieved Jan. 2005 from: http://www.richters.com/newdisplay.cgi?page=MagazineRack/Articles/WhatHappened.html&cart_id=977976.16253.

Rickert 2004. *Rickert's bee preservation site. Rickert's honey bee page.* Retrieved Apr. 2005 from: http://www.holoweb.com/cannon/insects.htm.

Ripa, A.K. 1993. *Introduction of the cowberry (Vaccinium vitis-idaea) into cultivation.* Aquilo Ser. Bot. 31:55-58.

Road Saver Plus® 2004. *Synopsis – relative effectiveness of road dust suppressants.* Retrieved June 2004 from: http://www.searsoil.com/comparison.htm.

Roberts, D.R. and Sutton, C.S. 1995. "Research in conifer somatic embryogenesis." In: *Research in Canada. Recent Progress in Forest Biotechnology in Canada.* PI-X-120. 1-9.

Rosenqvist, H., Roos, A., Ling, E. and Hektor, B. 2000. *Willow growers in Sweden.* Biomass and Bioenergy 18:137-145.

Rosenweig, R., Varilek, M., Janssen, J. 2002. *The emerging international greenhouse gas market.* Pew Center on Global Climate Change. (Online Version: http:www. pewclimate.org/docUploads/trading%2Epdf).

Rosnew, H. 2004. *Turning genetically engineered trees into toxic avengers.* New York Times. Published Aug. 3, 2004. Retrieved Aug. 2005 from: http://query.nytimes.com/gst/ health/article-page.html?res=9A04E3DB163CF930A3575BC0A9629C8B63.

Ross, N. 2004. *Carbon sinks, forest rises.* Alternatives Journal 30 (3): 8-12.

Ross, R. 1998. *The changing future of NTFP's.* In: Non-Timber Forest Products Workshop Proceedings, April 3-5, 1998. Inner Coast Natural Resource Centre, U. of Victoria, Alert Bay, B.C.

Saari, V. 1993. *Collection products of transmission-line corridors and their utilization possibilities.* Aquilo Ser. Bot. 31:47-54.

Samson, R., Girouard, P., Zan, C., Mehdi, B., Martin, R. and Henning, J. 1999. *The implications of growing short-rotation tree species for carbon sequestration in Canada, final report.* Prepared for The Joint Forest Sector Table/Sinks Table Afforestation #5, National Climate Change Process, Solicitation No: 23103-8-0253/N, Resource Efficient Agricultural Production (R.E.A.P.)-Canada, Apr. 1999. (Online Version: http://www.reap-canada.com/Reports/SRF.htm).

Saskatchewan Agriculture, Food and Rural Revitalization 2000. Special forest products. Retrieved Mar. 2003 from: http://www.agr.gov.sk.ca/docs/crops/norther_agriculture/ spforest.asp.

Saskatchewan Agrivision Corporation Inc. 2002. *The nutraceutical, functional food and dermaceutical industry: A SAC Inc. state of the industry fact sheet.* Retrieved Apr. 2004 from: http://www.agrivision.sk.ca/Archives-PDF/FactSheets/Nutraceutical.pdf.

Saskatchewan Environmental Society 2002. *Non-timber forest products: Economic development while sustaining our northern forests.* Saskatchewan Environmental Society, Saskatoon, Canada. (Online Version: http://www.environmentalsociety.ca/resources/ publications.html).

Saskatchewan Environmental Society 2002a. *The ecotourism incentive: economic development while preserving the environment.* Saskatchewan Environmental Society, Saskatoon, Canada. (Online Version: http://www.environmentalsociety.ca/issues/forests/ ecotourism.pdf).

Saskatechewan Nutraceutical Network 2004. *Ag-West Bio Inc.Publications.* Retrieved Nov. 2004 from: http://www.nutranet.org/.

Savage, M. 1995. *Pacific Northwest special forest products: an industry in transition.* Journal of Forestry 93(3):6-11.

Schlosser, W.E., Blatner, K.A. and Chapman, R.C. 1991. *Economic and marketing implications of special forest products harvest in the Coastal Pacific Northwest.* Western J. of Applied For. 6(3):67-72.

Schmidt, C.W. 1998. *Natural born killers.* Environmental health 2000 (news). Environ Health Perspect 106:A600-A603.

Scholz, V., Berg, W. and Kaulfuß, P. 1998. *Energy balance of solid biofuels.* J. Agric. Engng Res. 71:263-272.

Schooley, J. 2003. *Marketing and export of ginseng.* Ontario Ministry of Agriculture and Food. Ginseng Series. (Online Version: http://www.gov.on.ca/OMAFRA/english/crops/ facts/ginmkexp.htm).

Schor, J. 1994. *The evolution and development of biotechnology: A revolutionary force in American agriculture.* US Department of Agriculture Economic Research Service. Washington, D.C. (Online Version: http://www.accessexcellence.org/AB/BA/future_fuel. html).

Schut, J. 2002. *Fulfilling a dream of moldable wood.* Plastics Technology. Mar. 2002. Retrieved June 2005 from: http://www.plasticstechnology.com/articles/200203bib2.html

Scott, G.M., Akhtar, M. and Swaney, R.E. 1998. *Economic evaluation of biopulping.* 7th International Conference on Biotechnology in the Pulp and Paper Industry, Vancouver, BC, Canada, June 16-19. Retrieved Mar. 2003 from: http://www.fpl.fs.fed.us/ documents/pdf1998/ scott98d.pdf.

Scott Wolfe Management 2002. *Potential benifits of functional foods and nutraceuticals to the agri-food industry in Canada - final report.* Agriculture and Agri-Foods Canada, Food Bureau. Retrieved Mar. 2004 from: http://www.agr.gc.ca/food/nff/pdfdocs/ AgBenefit_Final_Report.pdf.

Sellman, S. 2002. *Xylitol – our sweet salvation.* Total Health, Sep/Oct. 2002 (Online Version: http://www.principalhealthnews.com/article/bellhowell/103285514).

Shin, H., Zeikus, J. and Jain, M. 2002. *Electrically enhanced fermentation by Clostridium thermocellum and Saccharomyces cerevisiae.* Appl. Microbiol. Biotechnol. 58:476-481.

Siitonen, J., Martikainen, P., Punttila, P. and Rauh, J. 2000. *Coarse woody debris and stand characteristics in mature managed and old-growth boreal mesic forests in southern Finland.* Forest Ecology and Management 128:211-225.

Silveron Health Products 2005. *Elk velvet antler and ginseng.* Retrieved July 2005 from: http://www.silveronhealthproducts.com/ginseng.

Simard, R.R., Lalande, R., Gagnon, B., Parent, G. and Parent, P. 1998. *Beneficial use of paper-mill residue compost in potato production.* The Composting Council of Canada. (Online Version: http://www.compost.org/cccPaperMillResidueCompost.htm).

Singh, P., 1993. *Research and management strategies for major tree diseases in Canada: synthesis. Part 1.* For. Chron. 151-162.

Skinner, B. 2002. *Fireweed.* US Fish and Wildlife Service, Arctic National Wildlife Refuge. Retrieved Mar. 2003 from: http://www.r7.fws.gov.nwr/artic/firewdra.html.

Sloan, E. 1999. *Hot! Hot! Hot! The top ten up-and-coming nutraceutical markets.* Nutraceuticals World, Mar./Apr. 1999. Retrieved Apr. 2004 from: http://www. nutraceuticalsworld.com/marapr 99-b7.htm.

Small Woodlands Program of British Columbia 2001. *A guide to agroforestry in BC.* National library of Canada Cataloguing in Publication Data. ISBN 0-7726-4618-X. (Online Version: http://www.woodlot.bc.ca/swp/).

Small, E. and Catling, P.M. 1999. *Canadian medicinal crops.* National Research Council of Canada Press, Ottawa, ON. 240 pp.

Smit, W.A. 2003. *Medicinal mushrooms.* The South African Journal of Natural Medicine. Issue 10.

Smith, R. and Cameron, S. 2003. *An introduction to ground hemlock (Taxus Canadensis).* Slide Presentation from 'Economic Opportunities in Non-timber Forest Products, Atlantic Region NTFP Workshop,' Feb. 1, 2003. Retrieved Mar. 2004 from: http://www. fallsbrookcentre.ca/webmain/programs/Forest/NTFP Web/Econ Opp Wkshop/ground hemlock.pdf.

Spillsorb 2003. *Spillsorb Canada company and product information. Retrieved* Feb. 2004 from: http://www.spillsorb.com/.

Stamets, P. 2005. *Mycofiltration: A novel approach for the bio-transformation of abandoned logging roads.* Retrieved Aug. 2005 from: http://www.fungi.com/mycotech/roadrestoration.html.

Stanosz, G.R., Kruger, E.L. and Isebrands, J.G. 1998. *A prototype hybrid poplar biofuel plantation in southern Wisconsin.* Poster presented at Bioenergy '98: Expanding bioenergy partnerships. Bioenergy '98: Expanding Bioenergy Partnerships, Madison, Wisconsin, Oct. 4-8, 1998.

Statistics Canada 1997. *Use of farmland, Canada and the provinces.* Historical Overview of Canadian Agriculture (data and analytical products: 1996 Census of Agriculture). Catalogue 93-358-XPB, Released Aug. 14, 1997. (Online Version: http://www.statcan.ca/english/Pgdb/phys10.htm).

Statistics Canada 1999. *Measuring the attractiveness of R&D tax incentives: Canada and major industrial countries.* Catalogue No. 88F0006XIB1999010. Retrieved Sept. 2005 from: http://www.statcan.ca/english/research/88F0006XIE/88F0006XIB1999010.pdf.

Statistics Canada 2001. *Table 13 – Hay and field crops, by province, census agricultural region (CAR) and census division (CD), 2001.* Retrieved Mar. 2004 from: http://www.statcan.ca/english/freepub/95F0301X1E/tables/html/Table13Can10.htm.

Statistics Canada 2002. *Energy statistics handbook, quarter II, 2002.* Catalogue no. 57-601-XIE, ISSN 1496-4600.

Statistics Canada 2003. *Waste management industry survey, business and government sectors 2000.* Minister of Industry, 2003, Catalogue no 16F0023XIE, ISSN 1701-5677, Ottawa. (Online Version: http://www.statcan.ca/english/freepub/16F0023XIE/16F0023XIE00001.pdf).

Statistics Canada 2003a. *Production and value of honey and maple - 2003.* Statistics Canada Agriculture Division. Cataloge No. 23-221-XIB (ISSBN: 1481-6229) Ottawa Canada Retrieved Sept. 2005 from: http://www.statcan.ca/english/freepub/23-221-XIB/23-221-XIB03000.pdf).

Statistics Canada 2004. *Land and freshwater area.* Canadian Statistics, Statistics Canada, date modified: 2004-02-20. Retrieved Feb. 2004 from: http://www.statcan.ca/english/Pgdb/phys01.htm.

Statistics Canada 2004a. *Canadian international merchandise trade.* Retrieved Apr. 2005 from: http://www.statcan.ca/english/services/.

Statistics Canada 2004b. *Fur statistics 2004 vol. 2, no.1.* Agriculture Division. Retrieved Feb. 2005 from: http://www.statcan.ca/english/freepub/23-013-XIE/23-013-XIE2004001.pdf.

Statistics Canada 2004c. *Fur production.* Retrieved Oct. 2004 from: http://www.statcan.ca/english/Pgdb/prim46.htm.

Stavins, R.N., K.R. Richards, K.R. 2005. *The cost of US forest-based carbon sequestration.* Pew Center on Global Climate Change. (Online Version: http://www.pewclimate.rog/docUploads/Sequest%5FFinal%2Epdf).

Sundance Health Products 2004. *Pure Canadian elk velvet.* Retrieved Nov. 2004 from: http://www.geocities.com.elkvelvet/.

Sylvestre, P., Veillette, A. and Cormier, E. 1999. *Demonstration of a reclamation technique for the primary and secondary sludge generated by pulp and paper mills.* St. Lawrence Technologies, Industrial Wastewater, Saint-Laurent Vision 2000, Environment Canada, Mar. 1999. (Online Version: http://dsp-psd.communication.gc.ca/Collection/En1-17-40-1999E.pdf).

Taticek, R.A., Moo-Young, M. and Legge, R.L. 1991. *The scale-up of plant cell culture: Engineering considerations.* Plant Cell Tissue Org. Cult: 139-158.

TDC Marketing and Management Consultation 2004. *Heating with wood.* Retrieved Apr. 2004 from: http://www.tdc.ca/wood.htm.

Tea Council of Canada 2001. *Green and herbal tea fuel growth of tea sales in Canada.* Market Trends. Tea Association of Canada. Toronto Ontario, May 15, 2001. Retrieved Apr. 2005 from: http://www.tea.ca/press-trends-green.asp?section=media.

Tedder, S., Mitchell, D. and Farran, R. 2000. *Seeing the forest beneath the trees: The social and economic potential of non-timber forest products and services in the Queen Charlotte Islands/Haisa Gwaii, south Moresby forest replacement account.* (Online Version: http://www.for.gov.bc.ca/ftp/Het/external/!publish/web/non_timber_forest_products/qcis mf~1.pdf).

Tembec 2004. *Ethanol (alcohol).* Retrieved May 2005 from: http://www.tembec.com/ DynamicPortal.

Thadani, R. 2001. "International Non-Timber Forest Product Issues." In: *Non-Timber Forest Products: Medicinal Herbs, Fungi, Edible Fruits and Nuts, and Other Natural Products from the Forest.* Emery. M.R. and McLain, R.J. (eds.). Food Products Press, New York. pp. 5-24.

Thomas, M.G. and Schumann, D. 1993. *Income opportunities in special forest products: self-help suggestions for rural entrepreneurs.* USDA Forest Service Agricultural Information Bulletin 666. Retrieved Jan. 2005 from: http://www.fpl.fs.fed.us/documents/usda/ agib666/aib66601/pdf.

Totten, M. 1999. *Getting it right: emerging markets for storing carbon in forests.* Forest Trends and World Resources Institute. Washington, D.C. (Online Version: http://www.forest-trends.org/documents/publications/FTCarbon.pdf).

Toyo Tire Canada Inc. 2003. *Tires - Pneus.* Retrieved Jan. 2003 from: http://www. toyocanada.com/products/microbit.html.

Tsang, E.W.T., Charest, P.J. and Sederoff, R.R. 1995. "Genetic transformation in conifers." In: *Research in Canada. Recent Progress in Forest Biotechnology in Canada.* PI-X-120. 16-28.

Turner, N. and Szczawinski, A. 1988. *Edible wild fruits and nuts of Canada.* Canada's Edible Wild Plant Series, Vol. 3. Published by Fitzhenry and Whiteside, Markham, ON. ISBN 0-660-00129-2. 212 pp.

Turner, N.J. 2001. *"Keeping it living": applications and relevance of traditional plant management in British Columbia to sustainable harvesting of non-timber forest products.* In: Forest Communities in the Third Millennium: Linking research, business and policy toward a sustainable non-timber forest product sector. (Online Version:_http://www. ncrs.fs.fed.us/pubs/gtr/other/gtr-nc217/).

Turner, N.J. and Cocksedge, W. 2001. "Aboriginal Use of Non-Timber Forest Products in Northwestern North America: Applications and Issues." In: *Non-Timber Forest Products: Medicinal Herbs, Fungi, Edible Fruits and Nuts, and Other Natural Products from the Forest.* Emery, M.R. and McLain, R.J. (eds.). Food Products Press, New York. pp. 31-58.

United Nations Framework Convention on Climate Change 2003. *Caring for climate. A guide to the climate change convention and its Kyoto protocol.* Climate Change Secretariat. (Online Version: http://unfccc.int/resource/docs/publications/caring_en.pdf).

United Nations Framework Convention on Climate Change 2003a. *Issues in the negotiating process: Land-use - land-use change under the Kyoto protocol.* (Online Version: http://unfccc.int/cop7/lulucf.html).

University of Alaska 2004. *Birch: white gold in the boreal forest.* University of Alaska, Fairbanks. School of Natural Resources and Agricultural Sciences. Agricultural and Forestry Experiment Station. AFES Misc. Publication MP-04-02. (Online Version: http://www.uaf.edu/salrm/afes/pubs/misc/MP_04_02.pdf).

University of British Columbia 2005. Overview. *Michael Smith laboratories.* Retrieved Aug. 2005 from: http://www.michaelsmith.ubc.ca/research/overview/.

University of Nebraska 2005. *Levulinic acid from corn: Ingredient in antifreeze.* University of Nebraska-Lincoln. Industrial Agricultural Products Centre. Biopolymers, biochemicals, biofuels, biopower. Retrieved June 2005 from: http://agproducts.unl.edu/antifrez.htm.

University of Nebraska 2005a. *Horticultural uses of polylactic acid.* Industrial Agricultural Products Centre: Biopolymers, biochemicals, biofuels, biopower. Retrieved June 2005 from: http://agproducts.unl.edu/plahort.htm.

Upper Lakes Environmental Research Network 2004. *Canada Yew-developing a new value added crop for northern Ontario.* Northern Ontario Heritage Fund Corporation (NOHFC) Retrieved Mar. 2004 from: http://ulern.on.ca/Whatsnew/whatsnew.html.

US Department of Agriculture 1996. *Biopulping.* Pacific NW pollution prevention research center. Forest Products Laboratory. Retrieved Aug. 2005 from: http://www.pprc.org/pprc/rpd/fedfund/usda/usda_fpl/biopulp.html.

US Department of Agriculture 1999. *Trees of north Idaho: Pacific Yew (Taxus brevifolia).* USDA Forest Service, Idaho Panhandle National Forests. Retrieved June 2005 from: http://www.fs.fed.us/ipnf/eco/yourforest/trees/yew.html.

US Department of Agriculture 2003. *How many ramps do you need for a festival?* United States Department of Agriculture. Forest Service. Southern Research Station. Retrieved Apr. 2005 from: http://www.srs.fs.usda.gov/about/newsrelease/nr_2003-06-20-ramps.htm.

US Department of Agriculture 2004. *National agroforestry center.* (Online Version: http://www.unl.edu/nac/).

US Department of Energy 1998. *Commercialization of the biofine technology for Levulinic acid production from paper sludge.* Forest Products, Project Fact Sheet. Office of Industrial Technologies. Energy Efficiency and Renewable Energy, Washington, DC 20585. Retrieved June 2005 from: http://www.eere.energy.gov/industry/forest/pdfs/biofine.pdf.

US Department of Energy 1999. *Manufacture of industrial chemicals from levulinic acid: a new feedstock for the chemicals industry: Inexpensive biomass material to be developed into a variety of products.* Chemical Project Fact Sheet. Industrial Technologies Program, Washington, DC 20585. Retrieved June 2005 from: http://www.eere.energy.gov/industry/forest/pdfs/levulinic.pdf.

US Department of Energy 1999a. *Production of succinic acid from wood wastes and plants: New bacteria will be used as a biocatalyst to produce succinic acid from biomass.* Office of Industrial Technologies, Energy Efficiency and Renewable Energy. Feb. 1999. Retrieved July 2004 from: http://www.oit.doe.gov/chemicals/factsheets/succnic.pdf.

US Department of Energy 2005. *Microbial genome program.* Office of Science. Retrieved Aug. 2005 from: http://www.doegenomes.org/.

US Environmental Protection Agency 1999. *Economic conversion of cellulose biomass to chemicals.* Green Chemistry Challenge, 1999 Small Business Award. Retrieved June 2005 from: http://www.epa.gov/gcc/sba99.html.

van Oosten, C. 2000. *Activities related to poplar and willow cultivation and utilization in Canada 1996-1999.* Report to the 21st Session of the International Poplar Commission, Portland, Oregon, US, September 24-28, 2000. Poplar Council of Canada, 49 pp. (Online Version: http://www.poplar.ca/IPCCan2000.pdf).

van Seters, A.P. 1995. "Forest based medicines in traditional and cosmopolitan health care." In: *Non-Wood Forest Products 11, Medicinal Plants for Conservation and Health Care.* Food and Agriculture Organization of the United Nations, 8 pp. Retrieved Feb. 2003 from: http://www.fao.org/docrep/W7261e/W7261e04.HTM.

Vennum, T. 1988. *Wild rice and the Ojibway people*. Minnesota Historical Society Press, St. Paul. 55101. ISBN # 0-87351-226-X. 357 pp.

Ville de Montréal 2004. *Public seeding program for wild leek in Québec*. Biodome de Montréal. (Online Version: http://www2.ville.montreal.qc.ca/biodome/bdm.htm).

Vink, E.T. H., Rabago, K.R., Glassner, D.A. and Gruber, P.R. 2003. *Applications of life cycle assessment to NatureWorks^{TM} polylactide (PLA) production*. Polymer Degradation and Stability 80:403-419.

von Hagen, B. and Fight, R.D. 1999. *Opportunities for conservation-based development of nontimber forest products in the pacific northwest*. USDA Forest Service General technical report PNW-GTR-473.

Watts, S.B. and Kozak, R.A. 2000. *Status of forestry related research in Canada, 1999/2000*. FORCAST / Canadian Forest Service, Ottawa, On. 20 pp.

Weisz, P.B. and Marshall, J.F. 1979. *High-grade fuels from biomass farming: potentials and constraints*. Science 206:24-29.

Wheeler, N.C. and Hehnen, M.T. 1993. *Taxol, A study in technology commercialization*. Journal of Forestry, October 1993:15-18.

White Earth Land Recovery Project 2004. *Threat to wild rice*. White Earth Land Recovery Project. 1-888-779-3577. Retrieved Apr. 2005 from: http://www.sacredearth.com/ethnobotany/news/WildRice.html.

Whole Foods Market 2004. *Essential oils*. Retrieved Nov. 2005 from: http://www.wholefoodsmarket.com/products/wholebody/aroma.html.

Wildlands League 1997. *Nurturing diversity through ecotourism. Forest diversity – community survival: new directions in Ontario's forests*. Fact Sheet #8. Retrieved Feb. 2005 from: http://www.wildlandsleague.org/fact8.pdf.

Wills, R.M. and Lipsey, R.G. 1999. *An economic strategy to develop non-timber forest products and services in British Columbia*. Forest Renewal BC. Project No. PA97538-ORE. Cognetics International Research Inc. Bowen Island, BC. 115 pp. (Online Version: http://www.sfp.forprod.vt.edu/pubs/ntfp_bc.pdf).

Wolf, E. and Wortman, D. 1992. *Pacific yew management on national forests: A biological and policy analysis*. The Northwest Environmental Journal 8:347-366.

Wolfe, S. 2002. *Functional foods and nutraceuticals: potential benefits of functional foods to the agri-food industry in Canada*. Agriculture and Agri-Food Canada. Retrieved Apr. 2004 from: http://www.agr.gc.ca/food/nff/agbenefits/agbenefits_e.html.

Woman Motorist 2003. *Mercedes settles on methanol for 2004 fuel cell cars*. Retrieved May 2005 from: http://womanmotorist.com/index.php/welcome.

Wood, S.M. and Layzell, D.B. 2003. *A Canadian biomass inventory: Feedstocks for bio-based energy*. BIOCAP Canada Foundation. Industry Canada, Contract #5006125 (Online Version: http://www.biocap.ca/images/pdfs/BIOCAP_Biomass_Inventory.pdf).

World Health Organization 2000. *General guidelines for methodologies on research and evaluation of traditional medicine*. World Health Organization, Geneva. (Online Version: http://www.who.int/medicines/library/trm/who-edm-trm-2000-1/who-edm-trm-2000-1.pdf).

Wyman, C.E. and Goodman, B.J. 1993. *Biotechnology for production of fuels, chemicals and materials from biomass*. Appl. Biochem. Biotechnol. 30/40:41-59.

Yarborough, D. 2003. *Maine's wild blueberry crop 2002*. Wild Blueberry Newsletter. University of Maine Cooperative Extension. Retrieved Mar. 2004 from: http://wildblueberries.maine.edu/Newsletters/wildbluenews0103.pdf.

Zasada, J.C., Gordon, A.G., Slaughter, C.W. and Duchesne, L.C. 1997. *Ecological considerations for the north American boreal forests*. Interim Repot IR-97-024/July,

International Institute for Applied Systems Analysis. Laxenburg, Austria, 63 pp. (Online Version: http://www.iiasa.ac.at/Publications/Documents/IR-97-024.pdf).

Zimmermann, T.R. 2002. *June bug, medicinal insect.* Wilderness Way Magazine. Volume 2, Issue 4 (Online Version: http://wwmag.net/Pages/junebug.htm).

Index